About the Authors

David Owen is the author of 19 fiction and nonfiction titles, including *Tasmanian Tiger* and *Shark: In peril in the sea*. He is also the author of the popular 'Pufferfish' detective series set in Tasmania. He is the Official Secretary of the Governor of Tasmania.

David Pemberton is a wildlife biologist and former manager of the Tasmanian Government's Save the Tasmanian Devil Program. He has published scientific papers on a wide variety of conservation challenges, including seabird and seal bycatch in fisheries, and is co-editor of *Saving the Tasmanian Devil,* and co-author of *Tasmanian Tiger* with David Owen.

'...a valuable resource for anyone with an interest in these beautiful creatures.' ***The Victorian Naturalist***

'Written in the conversational yet informative tone of the good naturalist-lecturer . . . this will be the go-to book on Tasmanian devils for the foreseeable future.' ***Booklist***

'Casting new light on the Tassie devil's aggro reputation, this comprehensive but engaging paperback highlights the value of the endangered species.' ***Australian Traveller***

'An exquisite introduction to the social and natural history of the devils.' ***EcoHealth Journal Consortium***

BASS STRAIT
TASMANIA
KING IS
FLINDERS IS
BADGER IS
CAPE BARREN IS
CLARKE IS
HUNTER IS
THREE HUMMOCK IS
ROBBINS IS
Stanley
Marrawah
Redpa
HWY
BASS
Trowutta
Arthur R
Arthur River
Temma
Wynyard
Burnie
NARAWNTAPU NAT. PARK
Devonport
HIGHWAY
Greens Beach
George Town
Tamar R
WATERHOUSE CONSERVATION AREA
Bridport
Cape Portland
Musselroe Bay
Mt William
Waterhouse
Pyengana
HWY
TASMAN
Launceston
BEN LOMOND NAT. PARK
Perth
MIDLAND
POATINA
HWY
ESK
HWY
DOUGLAS APSLEY
Waratah
Trowunna WILDLIFE PARK
Mole Creek
Cradle Mtn
MURCHISON
Pieman R

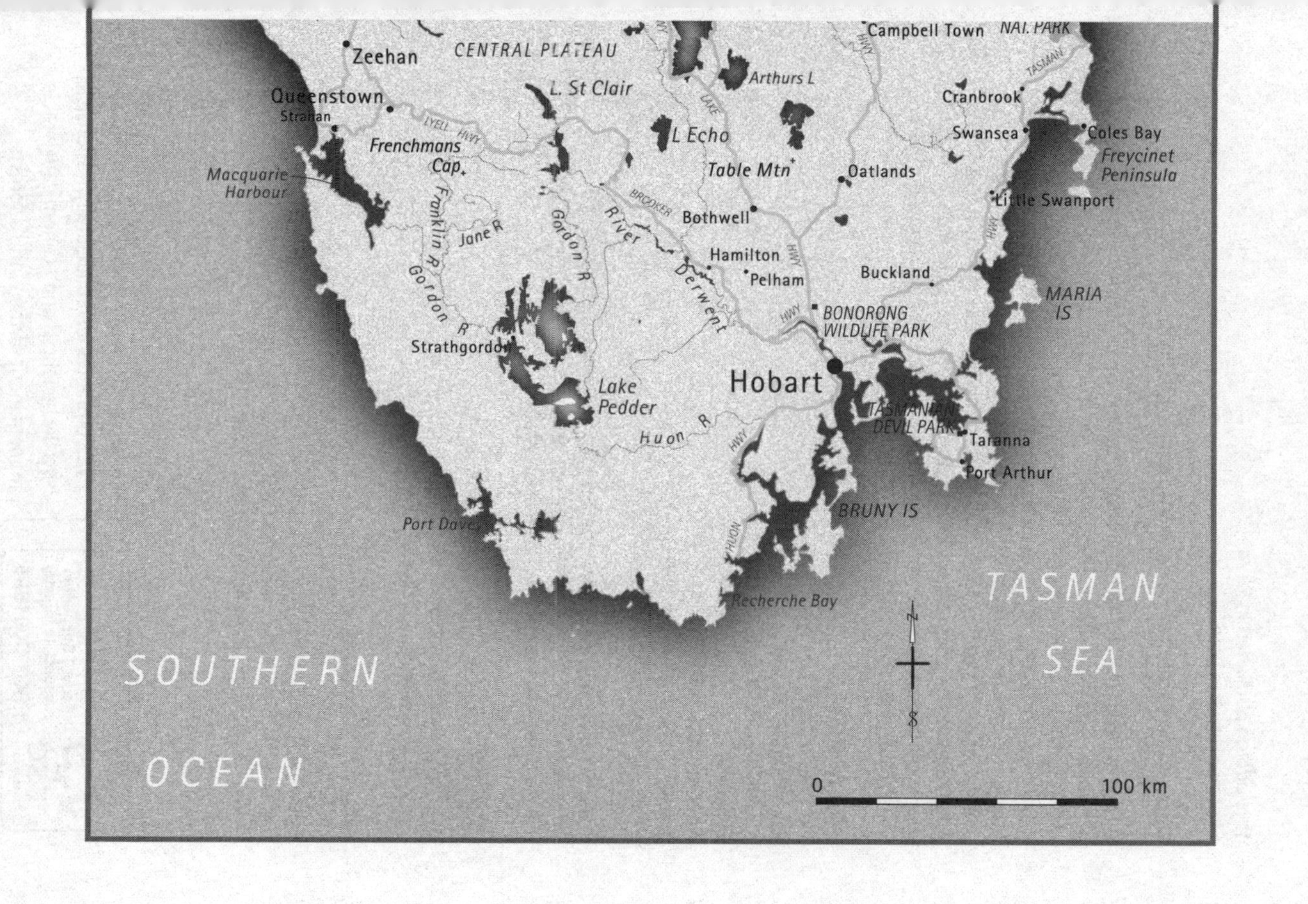
Zeehan
CENTRAL PLATEAU
Campbell Town
Queenstown
Strahan
L. St Clair
Arthurs L
Cranbrook
L Echo
Swansea
Coles Bay
Freycinet Peninsula
Frenchmans Cap
Macquarie Harbour
Table Mtn
Oatlands
Little Swanport
Franklin R
Jane R
Gordon R
River
Bothwell
Hamilton
Pelham
Buckland
Derwent
MARIA IS
BONORONG WILDLIFE PARK
Strathgordon
Hobart
Lake Pedder
Huon R
TASMANIAN DEVIL PARK
Taranna
Port Arthur
BRUNY IS
Port Davey
Recherche Bay
TASMAN SEA
SOUTHERN OCEAN
0
100 km

This edition published in 2024
First published in 2005

Allen & Unwin
Cammeraygal Country
83 Alexander Street
Crows Nest NSW 2065
Australia
Phone: (61 2) 8425 0100
Email: info@allenandunwin.com
Web: www.allenandunwin.com

Allen & Unwin acknowledges the Traditional Owners of the Country on which we live and work. We pay our respects to all Aboriginal and Torres Strait Islander Elders, past and present.

A catalogue record for this book is available from the National Library of Australia

ISBN 978 1 76147 040 0

Index by Puddingburn
Set in 11/14.5 pt Garamond 3 by Midland Typesetters, Australia
Printed and bound in Australia by the Opus Group

10 9 8 7 6 5 4 3 2 1

The paper in this book is FSC® certified. FSC® promotes environmentally responsible, socially beneficial and economically viable management of the world's forests.

DAVID OWEN AND
DAVID PEMBERTON

TASMANIAN DEVIL

SECOND EDITION

A deadly tale of survival

ALLEN&UNWIN

For Leisha, Hilton and Larry
D.O.

For Rosemary, Sam, Elsa and Ollie
and all the grandchildren
D.P.

CONTENTS

Acknowledgments xi
Introduction 1

1 Beelzebub's pup: a reappraisal of the Tasmanian devil 5
2 Evolution and extinction 28
3 Relationships in the wild 40
4 'Made for travelling rough': devil ecology 59
5 Devils and Europeans, 1803–1933 74
6 Demonising the Devil 94
7 Exploiting the Devil 104
8 Lure of the Devil 119
9 In the matter of the Society and the Board 126
10 From Antichrist to ambassador 145
11 In captivity 159
12 'The spinning animal from Tasmania' 169
13 Owning the devil: Tasmania and Warner Bros. 183
14 Devil Facial Tumour Disease 194

Notes 213
Select bibliography 229
Index 232

ACKNOWLEDGMENTS

Many organisations and individuals have assisted us in our work on this book, through permission to reproduce words or images and through personal communications.

In particular, and as will be seen throughout the book, Nick Mooney, former Tasmanian government wildlife biologist and a longtime advocate of protecting the island's fauna and unique environment, holds a special place in the story of the Tasmanian devil.

Likewise, zoologist and academic Dr Eric Guiler, devoted himself for over fifty years to championing the cause of Tasmania's wildlife, and his pioneering research into the Tasmanian devil is of seminal importance.

Dr Menna Jones and Androo Kelly are two other Tasmanians whose insights into its behaviour and ecology significantly increased our understanding of this rare marsupial carnivore.

The tragedy of devil facial tumour disease (DFTD) threatened the species with extinction. Through two decades of major local and international collaboration, this threat has been mitigated somewhat, but seems set to be part of the devil's life history.

Thanks in particular in the original preparation of this book are extended to Clare Hawkins, Billie Lazenby and Jason Wiersma.

Special thanks to Kathryn Medlock for her advice on the manuscript, to Christo Baars for permission to use his stunning devil images, and to Simon Bevilacqua for his enduring persistence in tracking the devil story. A big thank you also to Greg Irons and Jason Jacobi of Bonorong Wildlife Park, and also Aunty Patsy Cameron, a Tasmanian Elder.

We also wish to acknowledge and thank in our preparation of the book's original edition: Ian Bowring, Karen Gee, Catherine Taylor, Emma Cotter, Allen & Unwin; Bill Bleathman, Brian Looker, Belinda Bauer, Debbie Robertson, Tasmanian Museum and Art Gallery; Garry Bailey, Editor, The Mercury; Steven M. Fogelson, Senior Attorney, Warner Bros. Consumer Products Inc.; Patrick Medway, National President, Wildlife Preservation Society of Australia; Toren Virgis, Bonorong Wildlife Park; Elaine Kirchner, Fort Wayne Children's Zoo; Angela Anderson, Tasmanian Devil Park; Tony Marshall, Margaret Harman, State Library of Tasmania; Garth Wigston, Wigston's Lures; Judy K. Long, The Wolverine Foundation; Cheryl Vogt, Wilderness Safaris; Daniel J. Cox, Natural Exposures Inc.; Rob Giason, Tourism Tasmania; Ingrid Albion; Mike Archer; Bill Brown; Max Cameron; Donna Coleman; Rodney Croome; George Davis; Tim Dub; Rosemary Gales; Brian George; Lionel Grey; Lynda Guy; Lois Hall; Maureen Johnstone; Geoff King; Brendan McCrossen; Kate Mooney; Mike Myers; Jenny Nurse; Richard Perry; David Randall; Genene Randell; Steve Randell; Debbie Sadler; Alan Scott; Garry Sutton; John Teasdale; Russell Wheeldon; John R. Wilson; Stephen Wroe; Dave Watts; Janet Weaving.

David Owen and David Pemberton

INTRODUCTION

> All the visitors at Bronte Chalet leaned forward to see one of the world's most famous marsupials, the Tasmanian devil. His ears were pricked forward and there was almost a teddy-bear-like quality to his face as he moved towards his meal. Suddenly he turned and people reeled back in shock and horror. One whole side of his head was covered in a massive tumour like someone had stuck a slab of raw meat against his face. That was the last time we saw the devil nicknamed Phantom of the Opera alive . . . Mystical shocked me when her head first appeared as her face seemed to leap towards me. I then realised both lower jaws were covered in huge suppurating tumours. Amazingly her body condition was healthy with a big fat tail and she was even lactating and had young deposited in a nest somewhere. She was only three years old.
>
> INGRID ALBION, LAUDERDALE

Marrawah is a coastal township in the far northwest of Tasmania. Late in the twentieth century, a fifth-generation Marrawah farmer, the late Geoff King, chose to cease using his 830-acre property for cattle farming. Instead, he turned it into a wildlife sanctuary, specifically to protect the Tasmanian devil. Much of 'King's Run', as it was known, fronted wild, rocky coastline. A slow, bumpy ride through the scrub took visitors to his 'devil restaurant': an old, tiny, white wooden shack close to the surf, protected by huge slabs of quartzite and massive pebble banks thrown up by the Southern Ocean. The area has strong Aboriginal significance with an ethereal, otherworldly quality about it.

King and his visitors would chat, eat and drink in the basic but comfortable shack. Outside, near the window, a spot-lit wallaby carcass would be staked to the ground, a microphone alerting guests to the arriving devils. They generally turned up long after dark, hence the pleasure of socialising while waiting for them, briefly remote from civilisation. As often as not, it would be raining and blowing; the wind and pounding surf were a constant background and the smell of kelp was heavy and sweet. The magnified crunching of the devils, and their black-and-white, sharp but transient interactions with one another as they went about their complex feeding business, half in and half out of the discreet spotlighting, were vivid and magical to watch.

This book, too, gets as close as possible to the Tasmanian devil. It is written with great respect for the animal, which had seemed to represent a dynamic evolutionary success story on an ancient continent with a harsh environment. In its island home prior to European arrival in 1803 the devil presumably thrived. There are just a few known phonetic records of Aboriginal names for the devil (as written in the diaries of the controversial 'conciliator' George Augustus Robinson): *poirinnah* (Oyster Bay language); *tarrabah* (Bruny Island and Southeast language); *marripenner* (Ben Lomond language); *parloomerrer* (Northwest language), these being a number of the diverse languages recorded across the islands.

Now the devil facial tumour disease (DFTD), which so gravely afflicts the species, indicates that far from being a robust carnivore with no predator species to fear, it is highly vulnerable—a reminder of our limited understanding of the unpredictable natural world.

Christo Baars, a Dutch wildlife photographer, captured many striking images of Tasmanian devils. In 1996, near wukalina/Mount William in the state's far northeast, he chanced to photograph several devils with ghastly facial growths. Back in Hobart, Baars showed the photographs to the state government wildlife biologist Nick Mooney, who was horrified. While facial wounds, scars and abscesses are common in older devils, Mooney recognised something very different. Although cancer is a major cause of mortality in devils, it's usually internal. The discovery coincided with reports from farmers in the northeast, suggesting a drop in devil numbers. That part of the state had been associated with high concentrations of devils. Now, dead sheep and cattle lay uneaten in paddocks.

Cancer-like facial lumps and lesions on devils were not entirely unknown. In 1984 at wukalina/Mount William, David Pemberton had trapped a devil with an apparent facial tumour. In the early 1990s, there was an anecdotal report of a similar condition in the state's north. In 1999, Menna Jones, who was researching devils along Tasmania's east coast, observed some with tumours near Little Swanport, more than 250 kilometres south of where Baars had taken his original photographs. Two years later, she trapped three tumorous devils on the Freycinet Peninsula. Her subsequent monitoring there indicated that the peninsula's population was in serious decline. Along with Nick Mooney's trapping, it was becoming painfully obvious how the disease was spreading.

From an estimated peak of 150,000 devils in the mid-1990s, the state's numbers had dropped by at least one-third. This led to the first major step in tackling the disease: a specialists' workshop held in Launceston on 14 October 2003. Case by case, disease studies were presented at the workshop, helping to lead to the better understanding of devil cancers. The initial step was arriving at a clinical definition of DFTD, including this first 'official' case from 1995:

> History—A female Tasmanian Devil of unknown age was found at Greens Beach. Gross—very poor skin condition—hairless over much of surface, scabby, morocco leather-like patches, especially in the groin and flank. Subcutaneous lymph nodes enlarged and also scattered dermal and subcutaneous swellings. Necrotic reactionary lesion in the masseter muscle. The internal visceral organs appeared normal. Masses of cestodes (tapeworms) in the jejunum and moderate numbers in the ileum, plus there were ascarid-like nematodes mainly in the duodenal area. Histology—multifocal dermal leukosis with lymphosarcomatous infiltration of lymph nodes and in the periportal tissues of the liver and interstitially in the adrenal and also through skeletal muscle tissue. The lesion in the masseter muscle was due to heavy lymphosarcomatous infiltration of that muscle mass. This animal was affected by widespread lymphosarcomatous neoplasia. Diagnosis—Lymphosarcoma. Comment—There were occasional, but consistent findings of azurophilic intracytoplasmic material in some cells. Artefact or viral inclusions?[1]

Chapter 14 provides a comprehensive update on DFTD, now well into its third decade as the Tasmanian devil's uniquely unpleasant, complex and evolving fatal disease. But this book also demonstrates that the devil is robust. The world's largest marsupial carnivore remains reasonably safe in its island home, having endured a deadly tale of survival.

1

BEELZEBUB'S PUP: A REAPPRAISAL OF THE TASMANIAN DEVIL

> Over the years it got to be a war between Dawn and the devils as the stones, wires and other defences around the house foundations got bigger and bigger. Most years, however, the devils won. The growling would be heard and on inspection next morning they had dug their way through the stones and rocks to get back to their nest. Some nights the noise from the devils was amazing. Dawn had a big stick that she would bang on the lounge room floor to quieten them down . . .
>
> DEBBIE SADLER, ORIELTON

Good stories, no matter how unalike, share a tried and tested formula: intriguing setting; protagonist (good guy) and antagonist (bad guy); plot strength through mystery, drama and action; climax and resolution. In 1863 Morton Allport, a respected Hobart solicitor and naturalist, wrote a letter to his son Curzon describing a trip he had undertaken with a companion into Tasmania's alpine wilderness. An incidental paragraph of that letter exactly covers this formula, in small, slightly slanted handwriting:

> Before leaving Boviak Beach [setting], Packer [good guy] was considerably scared [drama] at meeting [action] what he called a Beelzebub's pup [mystery], in other words, a Tasmanian devil [bad guy], near to the camp but it made off [resolution] before the gun was ready [climax, suspended].[1]

The story of the Tasmanian devil is a remarkable one, surprising, controversial, funny, tragic. Nor has it been told before.

Few mammals have been so negatively named. In 1803, when a ragged boatload of English officers, sailors and convicts settled on the banks of the broad Derwent River, deep in the south of Tasmania, they wrongly assumed the island to be a physical extension of the east coast of New Holland, the name at that time for the Australian mainland. Their mistake was understandable, for in this new place were familiar eucalypt trees, kangaroos, wallabies and parrots. The devil, however, had been extinct on the mainland for centuries and so its vocalisations were unknown to these newcomers who, lying in their tents at night, listened nervously to the beast's alien shrieks and screams emanating from densely wooded mountains and valleys.

A reasonable hypothesis is that the new Vandemonians heard devils before seeing them, since the animals are nocturnal and rarely about during the day. And feared them: why else christen a small, 'night-screaming' carnivorous scavenger after the supreme embodiment of evil? On the other hand, there is something practical about the name. Beelzebub was Satan's first lieutenant, the prince of devils and 'lord of the flies'. Carcases, flies and Tasmanian devils have a lot in common.

Early written reports of the animal condemned it to persecution. It was incorrectly, though perhaps understandably, described as untameably savage, highly destructive to livestock and with such a fierce bite that ordinary-sized dogs were no match for it (even some small dogs like Jack Russells can kill devils). How to classify such a creature? The devil has had an array of taxonomic names, including the scary *Sarcophilus satanicus* (satanic meat lover) and *Diabolus ursinus* (diabolical bear). The most commonly accepted name is *Sarcophilus harrisii*, after the Deputy Surveyor General George Harris who in 1806 described and sketched the devil for the London Zoological Society. But some scientists have in recent times opted for *S. laniarius*, after mainland fossils so named in the 1830s by the French naturalists Georges Cuvier and Geoffroy Saint-Hilaire, and the English palaeontologist Richard Owen. To add to the uncertainty, there is also *S. moornaensis*, an even earlier mainland fossil, as well as another possible species nestled in time and size between *S. moornaensis* and the extant species.[2] On the other hand, in good Australian vernacular the devil might well be called the pied jumbuck-gobbler, *Gulpemdownus woollyturdii*.[3]

In 1830 the devil was singled out, along with the thylacine, as stock-destroying vermin to be eliminated through bounty schemes. Yet neither of these species was to blame for livestock losses, as shown by 80 years of bounty records painstakingly collected by Eric Guiler. The real culprits in the hard early years of the colony of Van Diemen's Land were poor management decisions and practices, and large packs of feral dogs. It has to be said, though, that the sight of a few devils tearing into a cast sheep or sick lamb does leave a strong impression.

Tasmanian bush myths perpetuate an incorrect fear that devils will attack and eat wounded or incapacitated bushwalkers. No such attack has ever been recorded. (Courtesy Nick Mooney)

And what of Packer's fear on Boviak Beach? It is true that devils will eat people, but only cadavers and only if the opportunity is there, such as finding a suicide or murder victim in the bush. On such occasions Tasmania Police forensic services invariably call upon experts, including former Nature Conservation Branch officers Nick Mooney and Mark Holdsworth.

There are, needless to say, Tasmanian devil bush myths, such as the couple hiking in the wilderness: one slips and becomes trapped under a fallen log, the other goes to get help, returns the next morning, and . . . only femur bones and boot-soles are left. In another, a drunk falls into a cattle trough and drowns with his arm hanging out, which gets eaten off.

But an element of caution is probably no bad thing. Alan Scott, manager of the Cameron family farm 'Kingston', at the foot of Ben Lomond, described the sprawling property as being 'in the middle of nowhere'.[4] (The late Major R. Cameron swore that in 1998 he saw a pair of thylacines on the property.) Scott said of the then newly rampant DFTD—that had struck hard in the region—that it was terrible no longer having devils about the place. Yet when he first began working on the property many years prior, and was required to do lone mustering on horseback in remote back paddocks, he feared the prospect of taking a fall, of being incapacitated far from help with night coming on. Indeed, Nick Mooney says he has come across this fear many times.

Devils are opportunistic feeders, not specialist predators. They eat a variety of foods.They occasionally kill reasonably sized prey such as sheep and wombats but forage more persistently for carrion, both vertebrate and invertebrate. While there are a few, unsubstantiated, reports of cooperative hunting, with one devil flushing prey and the other chasing it down—known as uncoordinated social hunting—the animal is overwhelmingly a solitary hunter. The devil's physique, stealth and ability to run quickly in short bursts make it a good night hunter, with wombats favoured for their fat content and their relative slowness. A devil is incapable of running down a bounding wallaby or sprinting rabbit. Its jaw strength and teeth have evolved to consume carrion, including tough gristle, skin and large bones.

Technically, if a sick lamb or wallaby is near death and a devil begins to eat it, that is an act of predation, not foraging, despite the devil having no role in bringing its 'prey' to that state. It is also technically correct to state that devils hunt tadpoles, yabbies and moths, which all feature in their diet. And if a devil opportunistically scents a nest of helpless baby quolls, native hens, wombats or, indeed, devils, and consumes them, that is predation, although the behaviour associated with the act is foraging.

That the devil is not a selective or timid eater is amply borne out, and not just by the antics of its tree-and-rock-chomping cartoon counterpart (which Warner Bros. brought to life in 1954 when there was hardly any available literature on devil behaviour). Items such as shoes regularly disappear off the verandahs of beach shacks and, if ever found again, have been well chewed. Devils love scavenging around rubbish dumps, but so do other opportunistic carnivores such as spotted-tailed quolls, cats, bears, hyaenas and foxes. Devils frequent beaches in search of dead fish and much else potentially edible deposited on the tideline. On one occasion at Geoff King's 'devil restaurant', during the day, Nick Mooney watched newly independent devils fossicking for and eating kelp maggots. David Pemberton observed similar feeding at remote south-west beach locations, including dead marine matter washed ashore.

David Randall, who worked as a ranger for many years in all parts of Tasmania, studied the native water rats of the Freycinet Peninsula on the east coast by trapping them with chunks of possum. One morning he found a sprung trap that a devil had subsequently broken into, forcing the wire apart to get at the bait and whatever creature was in there with it.

Garry Sutton, former ranger in charge of the Narawntapu National Park in northern Tasmania, once had to shoot an injured horse in the park. Because of its size and the park's public role, he used a front-end loader to dig a deep hole and bury the animal. Devils soon dug tunnels down through the sand to the decomposing corpse.

Staked out roadkill wallabies provide the bill of fare attracting devils to the island's scattering of devil restaurants. (Courtesy Nick Mooney)

One of Alan Scott's cows died giving birth. He left the corpse overnight and returned the next morning to see 'a devil coming out the backside'.[5] That's not unusual: the easiest way into a large animal is through the soft parts. And Guiler reported that he and a colleague 'found three devils sleeping off their feast inside the rib cage of a cow they were consuming'.[6]

The observed record for devils feeding simultaneously—also on a cow—is 22. This is a remarkable behavioural aspect of this generally solitary animal. It is also misunderstood behaviour, and one of the reasons why devils have such a bad reputation.

Far from being a free-for-all, communal devil feeding is structured and purposeful, and is properly described as ritualised behaviour. The screaming and apparent fighting is an elaborate combination and variety of vocalisations and postures by which order is maintained. The noises also act as a compass at night, alerting other devils in the area—just as daylight-circling vultures attract others—which saves them wasting energy looking for food. Smaller carcasses equal less noise. The perceived practice of eating 'everything'—because it disappears—is the result of individuals taking what they can and hiding with their share to consume it in peace.

Devils are the great hygienists of the Tasmanian bush and long ago extended that courtesy to farmers, eating their dead and sick livestock and in the process breaking the sheep tapeworm cycle, keeping the blowfly population down and relieving conscientious landowners of the need to bury dead stock. For these reasons there were proposals to introduce devils to Tasmania's Flinders Island and King Island, where roadkill wildlife smells and is unsightly, attracting adverse reactions from tourists. One problem with such a proposal is that the same danger would be posed to devils feeding by the roadside: speeding cars. A solution would be to heave roadkill into the bush, where it can be consumed in safety. Devil researcher Menna Jones and her volunteers did this along the road to Coles Bay on the Freycinet Peninsula almost every night during the years of her work there. The number of devils killed by cars decreased dramatically.

Communal feeding gives rise to the apparent paradox that this asocial animal indulges in a complex social ritual as often as every third or fourth night. David Parer, an internationally known film-maker specialising in wildlife documentaries, spent many years filming and observing devils and was well aware of the social nature of the species: 'We think of them as bad-tempered and vicious but watch them in the den and their family lives are not unlike a human life. There's playtime, squabbles, dinner time, discipline problems, teaching and learning'.[7]

Are Tasmanian devils dangerous? Of the many people working closely with devils—biologists, orphan carers, wildlife park employees—few would disagree that although individual animals have greatly varying personalities the species as a whole is timid. A wild devil trapped in a cage (though not a painful leg-hold trap or snare) will 'freeze' or become inert and won't struggle if carefully handled. Quolls, by contrast, bolt as soon as they can, and possums are notoriously difficult to handle. Guiler experimented empirically by putting devils and rats together in a small enclosure. The devils, he reported, were at times wary of the scurrying rats. If anyone knew them, Guiler did: 'most of the more than 7000 Tasmanian devils he handled were docile to the point of being lethargic and could be handled with ease'.[8]

On the other hand devil rage, though rare, is real. Early in 2005 two wildlife volunteers released a newly weaned devil from a trap. It turned and chased them so aggressively that they had to leap onto their car. More recently, a member of the Save the Tasmanian Devil Program was captured on video being chased in erratic circles by an angry devil. Devils can often be seen chasing one another in their wildlife park enclosures, and the same behaviour occurs in the wild during feeding bouts. Agitated/excited devils can perform a kind of spinning movement—is it pure coincidence that the famous Warner Brothers cartoon character Taz is a wildly spinning creature?

Devils scare easily and when startled will often shake. A screeching and biting devil acting purely out of fear will, however, if held firmly, become still. And they are sensitive. In 2004 both David Pemberton and his family and Nick and Kate Mooney hand-reared devils whose mothers had died from DFTD. Pemberton visited the Mooneys one Saturday, after

Juvenile devils display some of the characteristics of young domestic animals—inquisitiveness, playfulness. (Courtesy David Pemberton)

which, said Nick, their devil Eric, 'a charismatic charmer with a short fuse, came in behaving normally, until Kate fired up the vacuum cleaner—he changed completely and has remained one hundred per cent timid for days, hiding in dark corners. I suspect a combination of [lingering] smells from David's devils [on his clothes] and the size and noise of the cleaner may have told him that a large, dominant devil had entered the house'.[9]

Timid and sensitive, yes—yet in the popular imagination the devil has always been considered quite the opposite, as in this ludicrous 1917 description:

> Curiosity having been aroused as to why these ugly things received their highly suggestive name, it was stated that there can be little doubt that they deserved it. It is another case of ugliness going to the bone. Indeed, any virtues they possess are negative ones, and their vices are most positive. They are very savage, and have frequent fights among themselves, while

> they slay other creatures for the mere wanton lust of slaughter. When they attack anything, a member of their own tribe or any other species, they will practically tear it to pieces in sheer ruthlessness . . . During the day it is too sleepy to be otherwise than very stupid, but with the oncoming of covering darkness it displays a cunning and a cleverness inseparably connected in the human mind with the original owners of the despised name of devil.[10]

No less a scientist than Clive Lord, an eminent early Director of the Tasmanian Museum and Art Gallery, described the devil as 'exceedingly quarrelsome'[11] and later as 'of fierce disposition . . . It cannot be considered a pleasant animal to have much to do with'.[12] Devils were presumed to be scarce when Lord wrote about them

Errors of fact by professionals haven't helped the devil. This watercolour is by D. Colbron Pearse, who for a period in the 1950s was acting Director of the Tasmanian Museum and Art Gallery. The young are not accurate depictions of devils—rather, they are hybrids of spotted-tailed quolls and devils. (Courtesy Collection Tasmanian Museum and Art Gallery)

and, if they were, it is possible that he had relied on scanty and biased rural accounts for his observations, because he was wrong on all three counts.

Highly improbable is the link the Tasmanian devil supposedly had with the enduring American myth of the Jersey Devil. This creature allegedly came to exist in the eighteenth century in the swamps of south-eastern New Jersey, having a horse-like head, wings, cloven feet and thick tail. Sightings of it were regular, including one by Napoleon Bonaparte's brother Joseph while hunting there (Joseph lived in America between 1815 and 1832) and it was even seen in the company of a headless pirate. Joseph Bonaparte may have known of a strange new 'devil' animal because his sister-in-law, the Empress Josephine, kept a menagerie that included marsupials and a Tasmanian emu.

During one week in 1909, some 30 sightings of the Jersey Devil caused near-panic. The Smithsonian Institution speculated that it might be a Jurassic survivor, possibly a pterodactyl or paleosaurus which had survived in the region's limestone caves. More plausible, if that is the word, was that 'New York scientists thought it to be a marsupial carnivore'.[13]

But confusion arising out of words and myth pales beside reality on the ground. Tasmanian devils have been mercilessly persecuted. Nineteenth-century bounty hunting gave way to widespread strychnine poisoning in the early and middle decades of the twentieth century, with baits laid by farmers and also by trappers who made a living from possum and wallaby pelts. Devils also died from eating rabbits poisoned with arsenic. Snaring was a substantial business; in the 1923 season, for instance, 693 147 possums were snared and about half that number of wallabies.[14] Yet spotted-tailed quolls probably damaged more snared animals than devils did, because they had a greater ability to reach a carcass suspended above the ground.

Likewise poorly attached chicken wire won't protect poultry from hungry devils, but their disappearance from many rural areas has not meant that hens are now safe. Max Cameron, the owner of 'Kingston', bought twelve hens for a new property near Trowunna and put them in a sealed room overnight. A quoll got into the room through a drain and 'necked'—killed and blood-sucked—all of them.[15]

Yet the Tasmanian devil has been an easy target for so long that, like the ethereal Jersey Devil flying through the mists of its densely wooded swamps, truth and reality are secondary to established myth.

Devils can stink—if they have been in a trap or cage for an extended period and their coats become matted with excrement. Adults have a naturally musty, waxy odour, but young devils are as clean-smelling as puppies and kittens. The perception that the animals stink because they eat rotting meat is incorrect, and reflects a lack of first-hand experience.

Their famed jaw strength is very real—the equivalent of that of a dog four times their size, or, for their body mass, more powerful than a tiger's. Scientific analysis suggests that relative to its size the devil has the most powerful of all mammalian bites. The earliest steel wire traps used by Guiler proved useless, since some adult devils were able to chew their way through the thick wire. When trapping devils in the 1950s naturalist David Fleay was astounded to find a devil wearing a glistening yellow metal 'collar', until he recalled that two weeks previously a special composite foot-snare of brass wire and hemp set for a thylacine had been found sprung and bitten off.[16] Even so, devils are incapable of chewing through the biggest bones of large animals. And tough hide is very difficult to chew and ingest. Thus while wombat flesh and the prized fat is devoured, devils generally leave the backbone and adjoining skin of these rugged herbivores.

The devil's range of vocalisations is truly impressive. There are at least eleven distinct vocalisations, but describing them isn't easy. Writing in 1806, Harris described 'a sudden kind of snorting'.[17] One hundred and fifty years later Fleay heard that sound as 'wheezing coughs that sounded harshly like "Horace"',[18] while for the excitable *Reminiscences From the Melbourne Zoo* correspondent:

> If one could imagine a choir consisting of imps in the infernal regions, with every ear-splitting, brain-scratching sound grouped in hideous discords, the only earthly model that could be used as a guide would be a chorus from a company of Tasmanian devils.[19]

According to German conservationist Bernhard Grzimek:

> The Tasmanian devil which lived with us at Frankfurt Zoo for a number of years used to sing loudly and persistently when encouraged to. When cleaning out its cage, all we had to do was to stand in front of it and give the right note, and the animal would open its mouth and join in, keeping up the performance for quite a while. (I had earlier managed to get my wolves to sing in the same way.)[20]

Director of the Frankfurt Zoo, Grzimek devoted himself to wildlife conservation, particularly in Tanzania, and his influential 1959 documentary *Serengeti Shall Not Die* won an Oscar award. Eric Guiler also 'sang' with a number of the devils he kept at the University of Tasmania.

Mary Roberts owned Beaumaris Zoo in Hobart early in the twentieth century. She particularly loved devils and developed a close relationship with them; they eagerly responded to her calls.

It is not a myth that devils like water. Captive devils regularly splash around in their water pools and clearly enjoy it. On his trapping expedition Fleay witnessed a not uncommon sight: 'Another huge fellow . . . glared balefully from behind a shut [trap] door one morning and when I turned him loose he rushed for the river bank, dived into the icy current and swam strongly to the opposite side disappearing among the ferns'.[21] Nick Mooney has seen a devil swim powerfully across the fast-flowing, 50-metre wide Arthur River. Devils can also duck-dive, but not actually swim underwater.

Their speed on land has not been fully appreciated. The general perception is that, because they are short-legged and have an awkward-looking gait, they are incapable of running quickly. Guiler seemed to confirm this. 'It lopes along at about 3–4 kilometres per hour, but when chased it can make about 12 kilometres per hour for a short time. Several times we have caught devils by running after them when they have escaped while being handled.'[22]

Others have different opinions. David Randall and a wildlife officer friend, Reuben Hooper, were discussing the extent to which the devil could be an efficient chase-and-catch predator. They decided upon an empirical test. Randall released a devil which Hooper chased. The devil outsprinted Hooper, then suddenly stopped, turned and hissed at him, and he had to leap over the animal.

Artist and naturalist George Davis has had a lifelong interest in Tasmania's flora and wildlife. He believed that in their wild state, with no roadkill or livestock, devils have no choice but to hunt and catch, and he could testify to their fleetness of foot—he once chased one in a Land Rover. The Tasmanian Government's environment website states that devils have been clocked running on a flat road at 25 kmh for up to 1.5 kilometres,[23] and at Cradle Mountain Menna Jones clocked one running in her headlights at 35 kmh for 300 metres.

What is certain is that devils have great stamina. David Pemberton's fieldwork in the 1980s involved radio-tracking individuals throughout the night. A typical pattern emerged: an animal leaves its den after dark and, at a steady lope, uses tracks and forest edges to investigate known food-source areas. Bursts of speed are intermingled with periods of stillness, lasting up to half an hour. That pattern suggests an ambush predator. Towards the end of the night the devil sets up a rapid nonstop lope to return to its den around dawn. Its ability to travel a long way, at a good pace and quietly, is impressive and is a typical attribute of the polyphagous carnivores—those which according to dictionaries have an 'excessive' desire to eat. Devil speeds of greater than 10 kmh are common for extended periods through the night, three or four times per week.

A famed devil story that happens to be true relates to sheep in shearing sheds, and the problem likely to beset one if a leg slips through the flooring slats and becomes stuck. Lionel Grey, a cull shooter, once came across a sheep in a shed 'with the hock chewed off it',[24] as did Helen Gee, who farmed at Buckland. Sceptics wonder if this is another farmers' myth to tarnish the devil, asking why sheep would be left overnight in a shearing shed.

But they frequently are, being penned overnight for an early start in the morning. And because many farms have been cleared of even small stands of timber, devils perforce set up dens under buildings, including shearing sheds.

Older Tasmanian houses sometimes have devil dens in their foundations, having possibly been in use for more than a hundred years, with no one being aware of their presence, though smell and noise are usually the giveaway. Nick Mooney was often called to remove devils from under houses. His preferred method was to install a one-way cat flap at the den entrance once any juveniles were large enough that they were naturally emerging and using other, secondary dens.

He has had a few memorable den experiences. One couple reported devil pups under their house and wanted to know if they could be shifted. Mooney talked them out of it for that season. The husband was a school soccer coach and after one match he brought two teams' jumpers home to wash. He left them in a bag on the verandah overnight and they subsequently disappeared. He assumed they had been stolen. Mooney reckoned otherwise. Knowing where the devils were denned, under the kitchen floor, he popped a plank—and there were the jumpers. He recollects that they fished up about 30 using a wire. None had been chewed and after being washed were fine. They retrieved a number of other items as well, including a pillow.

At another house, a litter of young devils was attracted to a feather-filled doona being aired on a clothesline. They pulled it off the line and tried to drag it through a hole into the house's foundations. They managed to get most of it in before it burst, showering the foundations with feathers. Mooney recalls wet black noses with white feathers stuck on them, and plenty of devil sneezing. He bought the owner a new doona.

During 2004, David Pemberton and his partner Rosemary Gales hand-reared two devils, Donny and Clyde, who made their den under a bed in a spare room. They regularly took items of newly washed clothing to the den. By the time Donny and Clyde were moved to the wild, they had lined their den with the equivalent of three basket-loads of clean washing.

A fact that could be mistaken for a myth is the tendency of devils to all go to the toilet in the same spot. The use of communal latrines is not common among animals. Brown hyaenas and ratels (honey badgers), two other species associated with the devil through convergent evolution, also use communal latrines. They are instances of an apparently solitary animal engaging in at least chemical social interaction. Depending on population numbers, dozens of devils will defecate in one area—usually near a creek crossing or other water source—for reasons of communication barely understood, and further calling into question the 'solitary' tag. The same spot will be reused by a devil after an absence of a week or more, which implies a form of territoriality. Devil latrines could be described as community noticeboards; they may tell transients that a particular area is full, and they may tell competing males something about female availability. San Diego Zoo tests indicate that devils are able to recognise other individuals through communal latrine use. These latrines may even have an inter-species communication function: spotted-tailed quoll scats have been found at devil latrines.

Nick and Kate Mooney hand-reared many orphaned devils for rehabilitation and had little difficulty toilet-training them because of the innate behavioural tendency to use one latrine.

Devil scats are huge and in them, as befits an unfussy feeder, are to be found a great variety of objects. So big are they relative to the animal's size that they have often been cited as evidence of the continued existence of thylacines. An average scat is about 15 centimetres long, but they can be up to 25 centimetres long.

Baby and juvenile devils are cute, playful, mischievous—and noisy, especially during the night. They climb whatever they can and play games which involve ambushing, chasing and dragging one another by the ear. David Pemberton, while rearing orphans, has observed that juvenile devils use their tails to send a range of excitable and nervous signals, with the tail bent stiffly toward the ground and twitching energetically. (Raised tails in most animals are generally considered the demonstration of a highly excitable state.)

Yet devils would not make good pets. Even little ones have formidably sharp teeth and vice-like jaws. Above all, once weaned they generally become asocial, which is presumably why Aboriginal people, who quickly took to dogs after European settlement, did not keep them as pets. This hasn't prevented some Australian scientists suggesting that endangered marsupials be tested as pets, with a view to breeding them up. A report on ABC Radio's *PM* program began with host Mark Colvin introducing the topic this way: 'Imagine curling up in front of a winter fire with a Tasmanian devil at your feet, or an eastern quoll on your lap . . .'[25] Having said this hand-reared devils do like their comfort— some of the Mooneys' winter orphans would gather at the fireplace and wait for the fire to be lit. And some adult devils do have what we might call calm, sweet natures.

Their protected status has not prevented a number of US exotic pet websites advertising devils. The international trade in exotic living things is vast and much of it illegal. It would be surprising if devils didn't form part of it, because they would sell handsomely, thanks in part to the high profile of the Warner Bros. cartoon character Taz. They are easy to catch, feed and house. But Tasmania's rural population is small and interconnected and locals involved in such a trade would have to go about their cruel business with great caution.

An incident in Perth, Western Australia, in July 1997 appears to confirm that there is such a trade. As reported by CNN, a woman found:

> an unusual illegal immigrant hiding under her car: a Tasmanian Devil . . . The Department of Conservation and Land Management did a little checking around. There are 16 registered licensees in Western Australia who are permitted to keep Tasmanian Devils, and none of them was missing any. The department believes the animal was imported illegally and kept as a pet before escaping.[26]

US Navy aircraft carriers occasionally visit Hobart, and in one year in the 1990s strong rumours were about that a number of sailors with Tasmanian devil tattoos—the animal was their group mascot—swapped or attempted to swap handguns for live devils.

Devil experts are occasionally asked if the animals can interbreed with dogs, the unspoken reason being a desire to breed a presumably omnipotent fighting hound.

Cruelty and ignorance have hurt the devil in many ways. One or two individual farmers are believed to have had an annual kill rate of over 1000 a year, through strychnine poisoning, trained dogs and mass trapping. George Davis witnessed a particularly cruel method of killing them. A northern farmer placed a

220-litre water tank in the ground and ran a baited drop-plank over it, luring devils onto the plank which then tipped them into the tank, where they fought and ate one another.

An east coast farmer used to kill them by nailing a baited shark hook to a tree trunk, at a height that would hook the devil on tiptoes so that it couldn't escape and would die in agony. A head keeper at Bonorong Wildlife Park witnessed fifteen shot devils being thrown on a bonfire. A senior Parks and Wildlife officer was heard to say in the 1980s that while he would avoid a wombat on the road, devils were fair game.

In 1993 Mooney found 32 dead devils around poisoned sheep carcasses, near a popular trout fishing spot in the central highlands. All had had their saddles skinned off. It appeared to be a mass killing for perhaps a floor mat, and such a mat may well adorn a central highlands fishing shack.

In 1952 David Fleay wrote:

> Fur trappers who still carry on large scale operations during the winters of western Tasmania heartily dislike the snare-despoiling Devil, and often go to extreme lengths to rid a particular area of these animals before the season begins. An old pine shack below the frowning Frenchman Range is still known as the Devil's Camp—thanks to the pitiless work carried out by the first snarers there who poisoned and trapped the unfortunate carnivores so that their whitened bones lay in that vicinity for many years afterwards.[27]

The apparent capacity of the devil to survive in both 'plague' and dangerously low numbers, despite human interference, seems to be another of its remarkable features, but to believe so would be to perpetuate the myth that the animal will survive regardless of humans. There is no historical account of a devil with gross external tumours, which indicates that DFTD could be a 'new' disease and thus may be associated with human activity. An early twentieth-century decline—if it did happen—is more likely to have been linked to thylacine trapping, the snaring of possums and wallabies, and poisoning than to disease.

Can devil numbers sustain 'everything'? The question hinges at least in part on 'numbers'. Despite decades of research, devil population shifts defy easy explanation. George Davis recalls that, as a boy in Pelham during the early 1940s, the capture of a devil caused excitement because the creature was so rare. David Randall remembers them being very uncommon everywhere in the 1950s, and also in low numbers in the late 1960s. Yet by the early 1970s and again in the late 1980s, farmers in the east and northeast complained of 'plague numbers' threatening the sheep industry.

Devils don't always benefit from food provided by roadkill. (Courtesy Nick Mooney)

Interference with food supply may affect devils. Davis recalls night shoots when a bag of three or four wallabies was considered good. The introduction of spotlight shooting in the late 1960s, at the same time as a great increase in the amount of agricultural browsing land, meant that suddenly hundreds of carcasses were being dumped every night. More food meant more devils, and consequently a human-induced alteration to natural population dynamics.

Roads might be another influence on devils where roadkills are common, for instance near grain fields which are particularly attractive to wallabies. Do devils live in greater numbers near roads which offer up a steady supply of roadkill? It is impossible to know what influence human factors have on devil movements and especially their den sites, which are the critical factors in the home range location. If, over time, human activities have disrupted naturally occurring devil genetic dispersal patterns, the final outcome may be population chaos followed by extinction.

2

EVOLUTION AND EXTINCTION

> Late into the night with our little boat anchored just outside the weedline about thirty metres from shore we heard an ungodly commotion. Spotlight quickly activated to find a Tasmanian devil tearing open the tightly wrapped package of sandwiches which it had somehow managed to get out of an airtight lunch box. In the couple of minutes it took to start the outboard motor and push the boat through the weeds to shore the devil and complete contents of the lunch box were gone. The devil had obviously eaten in silence until it got to the sandwiches. It must have got frustrated with the plastic wrap hence the sudden noisy outburst.
>
> BRIAN GEORGE, SORELL

The Tasmanian devil has the distinction of being the world's largest living marsupial carnivore, though since an adult male devil seldom weighs more than 12 kilograms the species cannot be compared with dominant placental carnivores in other parts of the world, such as lions, tigers and wolves. Many factors, operating across millions of years, have resulted in the devil occupying this unique position.

Australia once formed part of the southern hemisphere supercontinent of Gondwana, together with what would become South America, Antarctica, Madagascar, New Zealand, India and Africa. While it is not known precisely how Australia's marsupials evolved, fragmentary fossil evidence suggests that lineages of protomarsupial stocks originating in South America journeyed

These prints on an iced-over creek demonstrate the unusual gait of the devil, which may have descended from an arboreal ancestor that hopped along branches. (Courtesy Nick Mooney)

across the then-temperate Antarctic landmass. Australia became a continent about 45 million years ago, floating free with a cargo of flora and fauna that would evolve in isolation until the continent collided with the Indonesian archipelago. That isolation enabled marsupials to diversify free of competition, but the 'floating laboratory' created competition of another kind, in the form of major climate changes brought about by variation in global weather patterns, Australia's northward movement towards the equator, and the Southern Ocean, wind and pressure changes created by that movement. Enormous inland seas and tropical forests came and went, periodically giving way to colder, drier conditions.

Although the continent had at times supported big mountain ranges, its general overall flatness provided little protection from the subantarctic winds that scoured away much of its surface. The remaining nutrient-poor soils, increasing surface salinity, decreasing rainfall, and extreme fluctuations between day-time heat and night-time cold, determined the long-term evolution of unique, often sparse, tree, plant and grass forms. Australia's herbivores developed accordingly. They became either nocturnal or crepuscular (active at dawn and dusk) browsers and grazers. There were none of the vast herds of grazing animals such as zebra and bison that developed on the lush grasslands of Africa and North America, for example, so there was limited scope for predators.

The devil's unknown ancestors may well have been tree-dwellers, eating insects, nectar, fruits and young leaves. As those creatures grew larger, their hind legs may have begun to operate in unison to cope with moving along branches, leading eventually to the hopping gait that is characteristic of many marsupials. This may even explain the devil's unusual gait.

The devil's specific lineage appears to be a result of dramatic climate change around the middle of the Miocene Epoch (16 million–5 million years ago). Australia had experienced a long period of warm, moist conditions. Inland seas and rivers dominated the continent and supported a great variety of animal, bird and aquatic life. Not surprisingly, many types of predators flourished in that period. But the rapid onset of the first of many ice ages changed that. Colder, drier conditions shrank the forests until, 'at its peak, far more than half of the continent became technically arid'.[1] Major extinctions resulted.

A few carnivores survived. Two were ancestors of the thylacine and the quoll genus, both of them hunters. It may be that specialist scavenging came to be an important niche, with the thylacine in particular ensuring a supply of carrion through its habit of selective feeding. This may be how the devil line arose. The species has no known earlier ancestry, unlike both the thylacine and quoll, which trace back at least 25 million years. The extinct species *Glaucodon ballaratensis* from the Pliocene (around 5 million–2 million years ago) is described as an 'intermediate form' between quoll and devil.[2] Although this suggests evolutionary experimentation in response to the increasingly dry environment, speculation based on fragmentary fossil evidence must be treated with care.

Australia's Miocene fossil record was considered poor until, in the early 1980s, the rich Riversleigh fossil deposits in remote northwestern Queensland were properly surveyed by Michael Archer, Suzanne Hand and Henk Godthelp and their team. At some 100 sites, huge numbers of limestone-encased fossils are preserved in ancient cave systems and waterways. According to the Australian Museum: 'Almost half of what we know about the evolution of Australian mammals in the last 30 million years comes from bones found at a single site in the Riversleigh fossil beds. Half of that was unearthed in one hour.'[3] And in the words of Sir David Attenborough: 'Their finds were so rich that on one occasion, they picked up bone and teeth from 30 different unknown species in as many minutes.'[4]

Riversleigh was granted World Heritage status, together with the much younger limestone fossil sites at Naracoorte Caves, in southeastern South Australia. There the devil is represented in the extraordinarily rich Fossil Chamber, a huge cave into which animals fell over a period of some 300 000 years, creating a gigantic cone of well-preserved bone deposits. Although Naracoorte and Riversleigh contain a wealth of information yet to be tapped, they have enabled a vivid reconstruction of Australia's relatively recent but mysterious age of marsupial megafauna.

These giant creatures established themselves as the continent became colder and more arid. They dominated during the most recent Ice Age into the Pleistocene Epoch but were then subject to rapid mass extinction, a process that began about 70 000 years ago and ended when the last of them died away about 20 000 years ago, though these time-spans are as controversial as the reasons put forward to explain the extinctions.

Sarcophilus laniarius is the devil species found in the Naracoorte Fossil Chamber. It was about 15 per cent larger than a modern devil, making its body mass about 50 per cent greater. But caution is necessary. 'The relationships between the living Tasmanian Devil and the larger Pleistocene form are in doubt . . . The living animal may either be a dwarfed version of

S. laniarius or possibly a different species that coexisted with the latter.'[5] It was the eminent nineteenth-century palaeontologist Richard Owen (who discovered and classified *S. laniarius*) who originally proposed the idea of different coexisting sizes, based on fossils discovered in the Wellington caves of New South Wales in 1877.

Giant devil bones have also been found in Queensland, Western Australia, New South Wales and Tasmania. The earliest fossil evidence is from the Fishermans Cliff locality in southwestern New South Wales, where the species is described as *S. moornaensis*. The first appearances of *S. laniarius* are in a fossil deposit in the eastern Darling Downs of southeastern Queensland, and in the Victoria Cave deposit in South Australia. Dating these sites is difficult, but the species certainly was present between 70 000 and 50 000 years ago. The Mammoth Cave site in Western Australia, where *S. laniarius* has also been found, may be as old as 70 000 years. The Devil's Lair cave deposit in Western Australia is dated at 11 000 to 30 000 years old and shows evidence of both devils and Aboriginal inhabitants. More recent deposits from the Holocene Epoch (the past 10 000–11 000 years) of *S. harrisii* are found throughout Australia, including on Flinders Island in Bass Strait.

A fragment of a megafaunal devil jaw in the Queen Victoria Museum and Art Gallery in Launceston is about 50 per cent larger than that of the extant species. It would no doubt have been a most efficient carrion eater because, like the present-day devil, it was designed to consume most parts of a carcass including bones.

It is also possible that the giant devil was a hunter as well as a scavenger. Another such hunter was *Megalania prisca* (ancient giant butcher), an enormous hunting goanna five or more metres long:

> The large skull was equipped with numerous recurved, scimitar-like teeth . . . Like its modern counterparts, *Megalania* probably scavenged from dead animals, but would have also been able to hunt and kill quite large prey . . . It would also have competed for prey with other large carnivores such as the Marsupial Lion, *Thylacoleo carnifex*.[6]

There is considerable debate about this latter statement. For a long time it was believed that *Megalania prisca*, along with the huge terrestrial crocodile *Quinkana fortirostrum* and the giant snake *Wonambi naracoortensis*, were the dominant Ice Age land predators—consigning the mammals to a lesser, inferior role. But according to University of Sydney palaeontologist Dr Stephen Wroe, 'the role of Australia's fossil reptiles has been exaggerated, while that of our marsupial carnivores has been undersold. The image of an incongruous continent dominated by reptiles in the Age of Mammals has real curiosity value, but it is a castle in the air'.[7] Wroe's assessment is based on an exhaustive re-examination of comparative weight and size estimates.

Thylacine evidence reinforces the possibility of different-sized devils coexisting as well as occupying a range of predator–scavenger niches. Seven or so genera of extinct thylacine have been discovered, dating back at least 25 million years, in a range of sizes, from that of a quoll (4 kilograms) up to about 18 kilograms. While the larger species were true hunting carnivores, the smaller species were likely to have foraged for reptiles, small mammals and insects. The relationship between devil and thylacine is close enough to infer similar evolutionary traits in the challenging Australian environment.

The demise of the megafauna was both swift and extensive: virtually everything in excess of about 40 kilograms became extinct. This meant more than 50 species. The carnivore family shrank arithmetically and literally, leaving only the devil, the quolls and a single thylacine species representing medium- to large-sized mammal predators. Indeed, all modern Australian marsupials are true survivors, reflecting 'the considerable evolutionary fine-tuning that has allowed them to cope with the drastically altered climates and escalating environmental stress of the last five million years'.[8] But was there something other than smaller size that spared them from the fate of the megafauna? Climate change, human influence, or a combination of the two, have all been proposed as the agent of the antipodean mass extinction.

Climate proponents argue that at the height of the most recent Ice Age, between 18 000 and 22 000 years ago, the Australian environment had become incapable of sustaining large herbivores. Their world shifted from being cold and dry to warm and dry; the bigger the animal, the less adaptable it was to rapid environmental change. Over a relatively short period of time, Australia's preponderance of rainforest gave way to open woodland, then to savannah, then to desert. Food and water ran out for all but the smaller, more robust creatures, and for some reason there was a fairly specific cut-off body size.

In the absence of irrefutable evidence, can the climate theory be tested? Modern Australia has long been under the influence of the so-called ENSO effect, being the combined influence of El Niño and the Southern Oscillation, the former disrupting the regular rainfall patterns of the latter through cooling of the upper layer of the southern Pacific Ocean. The result across eastern Australia in particular is drought, sustained over perhaps five years before returning warm sea currents create heavy rains and floods. It's unpredictable and harsh, and the continent's arid-adapted wildlife reflects that. But there are a few exceptions. The Daintree tropical rainforest system in far northeast Queensland supports abundant and complex populations of flora and fauna, while Tasmania, a significant area of which is Gondwanan remnant forest, supported the carnivorous devil, thylacine and eastern quoll after their mainland extinctions.

Proponents of the human interference theory believe that migrating waves of people slaughtered the megafauna to such an extent that they became extinct. This would have to have taken place well before the peak in late Ice Age climate aridity (to disprove climate as the culprit), and suggestions are that the megafauna began to be slaughtered about 46 000 years ago. This so-called Blitzkrieg hypothesis infers swift and rampant over-killing, as seemingly happened with the New Zealand moas and North America's mammoths and mastodons. A less bloodthirsty explanation is that regular slaughter for consumption, together with the introduction of fire management, which significantly altered grazing and browsing habitats, induced the same extinction result but over a far longer period. In this context, Stephen Wroe contends that a significant mid-Holocene increase in human land usage could have been a primary cause.[9]

The devil may have survived because of its comparatively small size and ability to become even smaller, through dwarfism. Or its place in the ecosystem may have been assured because it was capable of both hunting and scavenging. An intriguing, not unrelated, evolutionary question is: why did the thylacine survive but not the larger marsupial lion?

The subsequent extinction of the devil across mainland Australia may have been through multiple causes. The introduction of dingoes some 6000 years ago is generally considered to have been a key factor. Their predator–scavenger niches overlap; dingoes would undoubtedly forage for young devils; there have never been dingoes in Tasmania. And/or low genetic diversity in both devils and thylacines may have rendered them particularly vulnerable to diseases, as we see today with DFTD.

There may in addition have been a climatic factor. Devils thrive in temperate, well-covered Tasmania with its abundance of prey in a relatively compact area. Much of mainland Australia, on the other hand, has become an ever more arid and inhospitable environment since the devil survived the megafaunal extinction.

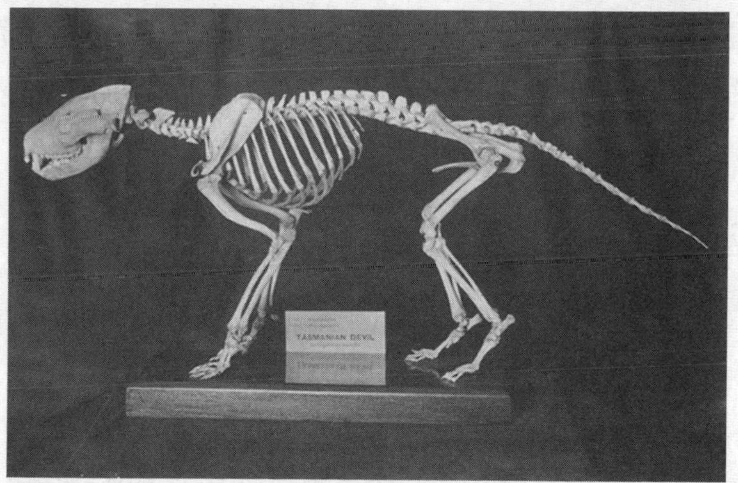

Skeleton of Sarcophilus harrisii, *the Tasmanian devil. (Courtesy Collection Tasmanian Museum and Art Gallery)*

Perhaps those conditions affected the mainland species over thousands of years until it was reduced to remnant populations in the east and southeast. (Then, and only then, might the devil have succumbed to the dingo.)

Megafauna-era butchering tools include scrapers of all kinds as well as axes, but it was not until some 10 000 years ago that the Aboriginal people became true hunters, with the invention of the boomerang and spear. It is suggested that prior to those technological advances, animal-taking must have been somewhat opportunistic. The famous Devil's Lair cave in south-west Western Australia, named for the extinct Tasmanian devil bones found in it, provides a clue.

Devil's Lair cave is one of the most important in Australian archaeology. By dating human occupation back some 45 000 years,[10] it confirms a much earlier human presence in the arid centre. Human markings on the walls may be the oldest on the continent. Cultural artefacts of bone and marl are also among the oldest known. Many extinct species are represented, but it appears that giant kangaroos were the primary food item, followed by wallabies and possums. If the devil was a food source, the scarcity of devil bones in the cave indicate either its rarity or a disinclination to catch and eat it. Of course, devils also live in caves.

There is, however, more recent evidence of the devil as a food source. Archaeological work at Victoria's Tower Hill Beach kitchen middens records 5000-year-old devil bones. Very few middens with devil bones have been found, but this did not stop one authority from declaring, 'the Aborigines knew how to hunt it, and they used it for food'.[11]

Writing in 1910 Fritz Noetling, Secretary of the Royal Society of Tasmania, cited a complete lack of evidence that

Tasmanian Aboriginal people consumed any of the marsupial predators or monotremes. 'It is undoubtedly very remarkable that even at the low state of civilisation represented by the Aborigines, human beings preferred the flesh of the herbivorous animals, and declined to eat that of the carnivorous.'[12] If this is true it may partly explain why in Tasmania devils and humans coexisted for tens of thousands of years prior to European settlement.

One of the greatest finds in Australian cultural history was made at Lake Nitchie, north of Wentworth on the Victoria–New South Wales border, in 1970. A male human skeleton, possibly 7000 years old, lay in a shallow grave. Unusually tall, he was wearing a necklace of 178 pierced Tasmanian devil teeth, collected from at least 47 animals. It has been speculated that the necklace indicates a dwindling population of *Sarcophilus*, and that it was considerably older than the skeleton. Archaeologist Josephine Flood went further: 'Indeed, if such necklaces were common, it is not surprising that Tasmanian devils became extinct'.[13] It is a startling suggestion, that the animal may have been hunted to extinction for its teeth.

On the other hand, the necklace is one of very few known to exist and required great labour to produce; this suggests it may have been of major significance. It is tantalising to speculate that the devil may therefore have held a special place in at least some societies of the distant past.

3

RELATIONSHIPS IN THE WILD

> I opened the tent zip and stood scanning the area until the beam came to rest on a large full grown Tassie devil looking straight at me only ten metres away. Closer inspection revealed a shiny object (my bloody fork!) hanging out of its mouth. Since I needed that fork more than he did, I charged the devil who dropped the fork and bolted into the bush.
>
> BRENDAN McCROSSEN, MIENA

Tasmania's devil is twice lucky, having escaped the ancient fate of its mainland counterpart and the overt consequences of European settlement, which in a little over two hundred years has accounted for the extinction of many of the Australian continent's mammal species. Most infamously, the thylacine, the Tasmanian tiger, was hunted during the nineteenth century as a supposed threat to the island's sheep industry. It has not been seen for over 70 years and in 1986 was declared officially extinct. The devil has replaced the thylacine as the island's largest marsupial predator, but because devils are also reliant on scavenging,

Tasmania no longer has a specialist cursorial (free-running) native terrestrial predator.

The Tasmanian tiger, a large pursuit predator, and the Tasmanian devil, a medium-sized ambush predator and scavenger, shared more than related names: their relationship in the wild was close and complex. Devils were probably preyed on by thylacines, but also benefited from the uneaten parts of the thylacines' prey. Being foragers, devils undoubtedly ate denned thylacine cubs, should they come across them unprotected. As thylacines became rarer, such incidental predation may even have hastened their demise. Old thylacines encroached on the devil's niche by scavenging. There may also have been competition for dens, given the preference of both species for caves, burrows and grass sags.

Anatomically the thylacine is dog-like (a good example of convergent evolution), is considerably more streamlined than the squat, stout devil, and properly described as a cursorial predator. Yet dentition studies carried out by Menna Jones confirm that devils and thylacines competed directly, their teeth demonstrating a significant niche overlap. Although the devil is only about half the weight of a thylacine, it is by comparison heavy-bodied and, with its speed over a short distance and powerful bite and forepaw grip, capable of bringing down prey larger than itself. She cites cases of devils attacking adult wombats of up to 30 kilograms.[1] Thylacines, however, show much less tooth breakage than devils, meaning less bone-eating. Thus, while the devil 'has a highly carnivorous dentition and trophic adaptions for bone consumption . . . The thylacine groups with the canids. Their molar teeth are intermediate in grinding and slicing functions and are quite slender, with no indications of adaption for bone consumption'.[2]

The devil's comparatively greater tooth and associated jaw muscle strength leads Jones to conclude that 'the role of top predator in the Tasmanian ecosystem was, at the least, shared equally between thylacines and devils'.[3] Although the concept of 'sharing and competing' may seem to be at odds with itself, there are successful examples elsewhere. Jaguars and pumas are roughly the same size as each other, but heavy-bodied jaguars take heavy prey such as peccaries, while the lighter-built pumas prey on smaller animals such as antelope. Jones speculates that the devil may have had a slight edge over the thylacine in taking heavy wombats.

Taxonomically, devils and thylacines are not that closely related, but their similarities and their greater differences provide insight into how evolutionary fine-tuning allowed them to coexist closely. They have in common distinctive markings:

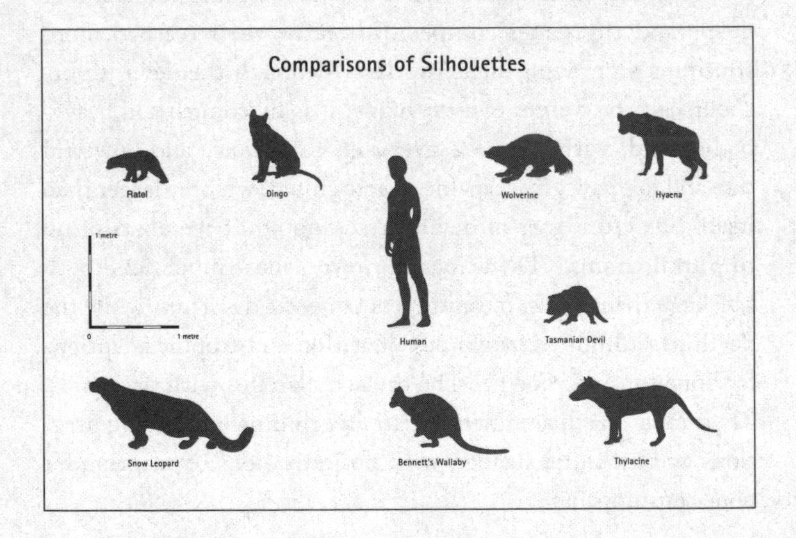

bold stripes, bold patches, which have camouflage, physiological and behavioural functions. Devil markings are important during feeding, the pure white flashes standing out at night in close interactions. The markings of both species are an indication of activity concentrated at dawn and dusk, less often during the day.

R.F. Ewer, a carnivore specialist in hyaenas and devils, speculated that a prototype/ideal canid would have both white markings and stripes to accentuate behavioural postures. While stripes aid in camouflage and possibly individual identification, white markings 'may serve to direct bites to relatively non-vulnerable areas'.[4] The thylacine had stripes, the devil has white patches. Where devil agonistic encounters result in bites they are typically on or near the rump where white markings are located (although the white chest-flash seems to play a role in initiating an encounter). Knowledge of devil marking is extremely limited. White flashes range from pronounced to marginal, with an estimated 13 per cent of animals being melanic, that is, all black. (DFTD appears to have had an effect on white splashes, increasing them, which could be an indicator of decreased genetic variety.)

The thick, largely non-prehensile tails of both species store fat. The devil's tail is important in physiology, locomotion and social behaviour. During high-speed motion it acts as a counterbalance. The tail is also a climbing aid.

The jaw gape of both species is wide, at 75–80 degrees, although for different reasons: thylacines used their gape to seize and suffocate or crush prey, while the gape and powerful teeth of the devil enable it to tear and gulp large lumps of food in a competitive manner, as well as to crush bones in order to consume them. Jaw gape is also a visual threat or defensive signal.

The animals' differences are pronounced. The devil's blackness is a sure asset for a small nocturnal creature; the thylacine's fawn or tan colouring shows a functional similarity to placental hunters such as wolves and wild dogs which hunt by day. An adult thylacine is about twice the weight and size of a devil. Anatomically it is considerably more streamlined than the squat, stout devil. This is because it is a cursorial predator, selecting its prey—generally a wallaby—and pursuing it relentlessly. Many early accounts refer to the thylacine's unhurried, dogged pursuit of prey, wearing its victim down through exhaustion, though it was undoubtedly capable of sharp speed over a short distance.

There was a long-held view that the thylacine was fussy and selective, consuming only the heart, kidneys and vascular tissue of its freshly killed prey, while the devil was a rapacious carrion eater. Clive Lord, director of the Tasmanian Museum in the 1920s, wrote:

> One or more Tasmanian devils will often follow a thylacine on its hunting excursions. The thylacine will kill a wallaby or other small animal, select a few choice morsels, and pass on. The devils will carry on the feast and consume the remnants, bones and all.[5]

Statements like this unwittingly consigned the two species to a strictly hierarchical relationship, and it is only in recent decades, through scientific studies, that the devil's predatory abilities have been recognised—and the likelihood that tigers and devils coexisted in a robust relationship.

Is the devil unique or can it be likened to non-marsupial mammals? Convergent evolution results when unrelated species in unrelated environments evolve similar adaptations because

Wolverines are sometimes referred to as the devil of the north. Their powerful teeth and jaws are adapted for chewing frozen carrion and crunching bone. (Courtesy Daniel J. Cox, Natural Exposures Inc.)

they occupy similar niches. Understanding and appreciation of the Tasmanian devil will be enhanced by finding convergent 'relatives' elsewhere. There are three good examples: the northern hemisphere wolverine (*Gulo gulo*), the southern hemisphere ratel (*Mellivora capensis*) and the hyaenas (striped, *Hyaena hyaena*; brown, *H. brunnea*; spotted, *Crocuta crocuta*).

Wolverines and ratels belong to the Mustelidae, the weasel family, which includes weasels, minks, polecats, otters, badgers and skunks.

Historically the wolverine has a broad circumpolar range, taking in Russia, the Scandinavian countries and North America. Despite the size of its range, in the words of the Wolverine Foundation, the umbrella organisation devoted to researching and protecting it, 'Even today, the wolverine remains largely a

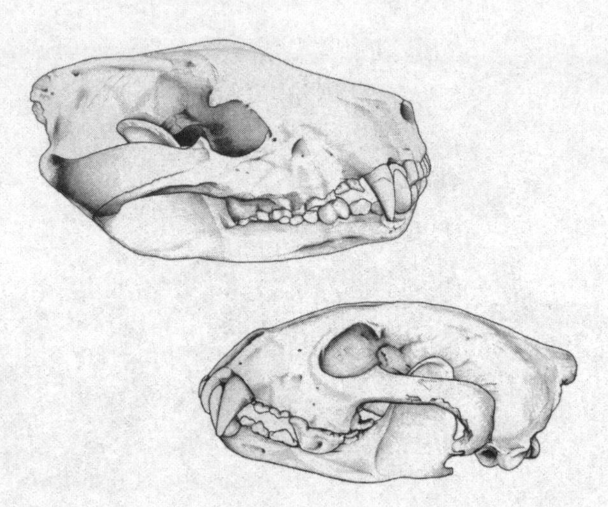

Comparative skull drawings of a Tasmanian devil (top) and a wolverine (bottom), showing the similarities of a robust carnivore skull. (Ian Faulkner)

mystery . . . one of the least understood and most fascinating creatures on earth'.[6]

In the United States, wolverines were once found as far south as California but appear to be confined now to Idaho and Montana, although there have been recent sightings in the Rocky Mountains states. Their Canadian range is also shrinking. The wolverine occupies a predator–scavenger niche one level down from the top predators, which in its range include wolves, bears, mountain lions and lynxes. This is not dissimilar to the Tasmanian devil–thylacine relationship.

Devils and wolverines have heavy builds, short powerful limbs, small round ears, weak eyesight, an excellent sense of smell and large heads to support their powerful jaws. (A wolverine skull in David Pemberton's office in the Tasmanian Museum and Art

Gallery is often mistaken for a devil skull.) Both animals have white neck and throat patches, and are occasional tree climbers—juvenile devils being more agile than adults, while generally wolverines are 'not considered to display arboreal behaviour'.[7]

Devils and wolverines are habitually described as nocturnal animals, but both species can be active during daylight. They are both regularly described as bear-like, due particularly to the shape of the head, the small eyes and round ears, broad chest when upright, glossy coat and pronounced claws. One of the Wolverine Foundation's Frequently Asked Questions is whether the Tasmanian devil is a biological relative of the wolverine. ('No . . . They may resemble each other physically, however they are distinctly different.')[8]

Adult wolverines typically weigh 2–3 kilograms more than adult devils, much of which is fat for insulation, while their fur is long and thick and their pads broad for travelling in snow. Wolverines have low density distribution; devil densities vary from low to very high. Wolverines have an inbuilt fearlessness and will not hesitate to attack if threatened, whereas devils generally are timid. Wolverines use strong scents to mark their territory. Devils often mark territory with an anal drag which spreads scent paste. Wolverines, as solitary animals, are quiet, as are solitary devils, though both have a range of social vocalisations.

Like the devil, the wolverine's food habits 'are weighted to scavenging'.[9] Wolverines are unfussy, opportunistic eaters and will cover large amounts of territory in search of food. They are known to attack incapacitated animals much larger than themselves but are generally recognised as voracious scavengers, so much so that early North American settlers nicknamed them

Gluttons. Eggs, insects, birds, rodents, squirrels and hares all form part of their diet.

Convergent evolution is strongly evident in the reason for both devils and wolverines having very powerful jaws: the wolverine is a specialist bone-crusher, capable of crunching through an elk or moose femur for the valuable marrow. It also chews frozen meat.

Both have been hunted for their persistent, opportunistic preying on animals trapped for their fur—wallabies and possums in Tasmania, mink and marten in the northern hemisphere. And just as the devil had a bounty placed on it for supposedly killing lambs in Tasmania's early days of white settlement, wolverines continue to be bounty-hunted across Scandinavia for their predation upon domestic reindeer and sheep.

Again, not unlike the devil, the wolverine is often regarded as a nuisance or worse, not least for its powerful, rank chemical secretions. In Native American folklore the wolverine is an ambivalent hero-trickster and a link to the spirit world. Interestingly, in North America it is sometimes called the Indian devil, and a 2002 video produced by the Wolverine Foundation is entitled *Wolverine: Devil of the North?*

The ratel, despite its common name of honey badger, is no longer classed in the badger sub-family. True badgers tend to be omnivores; ratels are predator–scavengers with a much greater tendency to carnivory. Hence, 'In 1902 it was transferred to the Mustelidae on the basis of skull morphology and teeth [and] in 1912 a kinship with the wolverine *Gulo gulo* was suggested.'[10]

Ratels occur throughout Africa—parts of the Sahara excepted—the Middle East and India. Not surprisingly, given that vast distribution, they are adapted to many forms of habitat,

The ratel, like the devil, is both a predator and scavenger. They have very powerful jaws. (Courtesy Mike Myers, Wilderness Safaris)

from dense wet rainforest to semi-arid desert and sub-alpine heights. Devils showed a similar widespread distribution across the Australian mainland, and they occupy all parts of Tasmania.

The ratel's lower body is black, the upper body light, although colouration varies according to habitat; they are paler in more arid regions. It is not inconceivable that mainland desert devils may have displayed some degree of colour adaptation.

Devils and ratels are thickset and low to the ground; ratels weigh slightly more than devils and have a proportionately longer body. The ratel has 'a massive head with a thick skull'.[11] The jaw is powerful, though unlike the devil's this may not be just for hunting and eating but also for defence: although considered 'shy and retiring',[12] it has a well-known propensity for aggression against animals much larger than itself.

The devil's peculiar lope is one of its most distinctive features, matched, however, by that of the ratel, which has a 'slow, rather bow-legged lumbering gait that sometimes increases to a clumsy gallop'.[13] Yet both, the ratel especially, are capable of running at considerable speed.

Convergent evolution is particularly evident in the diets of these unrelated, small, tough, nocturnal nomads. Devils eat 'everything'—from tadpoles to dead cows and horses—and one of the first surveys of ratels in the wild, undertaken in the late 1990s in the Kalahari Desert by zoologists Keith and Colleen Bigg, found likewise; ratels 'proved to be great opportunists, eating a range of 61 different species . . . food as small as social and solitary bee larvae, geckoes, scorpions, rodents and snakes to larger prey including springhares . . . birds and the juveniles of jackals . . . wildcat . . . fox'.[14]

Devils are closest of all to hyaenas, in particular the brown hyaena, which tends towards nocturnalism and solitariness, whereas spotted and striped hyaenas live in social clans of up to 80 with a rigid female-dominated hierarchy. Hyaenas and devils have a number of similarities: a shuffling lope resulting from a powerful forebody (hyaenas' rear legs are actually shorter than the forelegs); an ability to consume up to a third of their body weight at one feed, whether carrion or fresh kill; like the devil, brown hyaenas regularly forage on beaches (their range takes in the Western Cape, Namibia and Angola); hyaenas use communal latrines for social purposes; the vocalisations of hyaenas—screams, giggles, whoops, growls and snarls—match or exceed the devil's in range and complexity.

Hyaenas have a large sagittal crest on top of the skull for muscle attachment, giving the jaws great power—they are able to work their way through large bones.

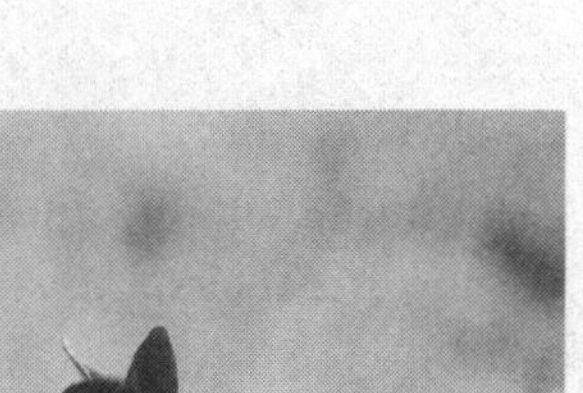

Convergent evolution is evident in the range of similarities between the brown hyaena and the Tasmanian devil, including a tendency towards nocturnalism and solitariness. (Courtesy Mike Myers, Wilderness Safaris)

Ignorance and superstition branded the hyaena a cowardly scavenger. Among its supposedly demonic attributes was an ability to change sex at will. The reality is that male and female hyaena genitalia are very similar, because females have a high testosterone count. Eric Guiler, intriguingly, claimed to have witnessed consecutive hermaphroditism—sex reversal—in a number of captured devils. David Pemberton and Nick Mooney, trapping devils at Granville Harbour in 2004, observed a devil with a non-functional pouch and scrotum.

One of the hyaena's practical functions, disposing of human corpses, may well have led many Africans to regard it with unease. There are still some remote tribes of Maasai and Karamajong, however, for whom this method of corpse disposal is delivery of the individual's spirit to the afterlife. The hyaena acts as a link: 'Every year since the 16th century, to mark the birth of Prophet Mohammad (PBUH), the city's residents offer the hyenas porridge mixed with butter and goat meat on the "Hakim Mountain", outside the city, believed to be a holy site because the ancient Muslim leaders of the city are buried there.'[15]

This is an interesting echo of the wolverine's link to the spirit world in Native American mythology. In pre-European Tasmania, might the devil have had such a relationship with the Indigenous humans?

In a role not dissimilar to that historically attributed to devils—that they followed thylacines and ate the remains of their prey—hyaenas were incorrectly portrayed as bickering scavengers cleaning up after lion kills. Although scavenging is important, all three species of hyaena are active, highly successful pursuit predators. And like devils, they will opportunistically eat their own.

Despite the relationship between the Tasmanian devil and the thylacine, the devil is taxonomically closer to the other members of the Dasyuridae family, the quolls and the tiny mice-like marsupials—dibblers, antechinuses, kowaris, mulgaras, kalutas, phascogales, planigales, ningauis, dunnarts and kultarrs. (The more distantly related numbats, bandicoots, bilbies and the marsupial mole make up the rest of Australia's sub-order of carnivorous marsupials.)

Quolls are well covered with spotted fur, have long tails, pointed facial features and sharp teeth. Two of the four species are found in Tasmania, the abundant eastern quoll (*Dasyurus viverrinus*, once called the native cat) and the larger, less common spotted-tailed quoll (*D. maculatus*, tiger cat), which weighs up to 7 kilograms. Both species were once common across the mainland, but the eastern quoll is extinct there now and its larger cousin reduced to rump populations. Quolls are excellent hunters and prey on many invertebrates, reptiles, rodents, possums and small macropods, the spotted-tail quoll also an efficient hunter of small wallabies. Spotted-tailed quolls climb well, enabling birds and sugar gliders to be included in their prey. But carrion also forms an important part of the diet, while rubbish-dump scavenging, poultry raiding, corbey grubs and fruit all add to an impressively varied diet.

The great mammalogist John Eisenberg visited Tasmania in 1990 on sabbatical. He had recently published *The Mammalian Radiations*, the most comprehensive summary of mammal evolution to date. Earlier in his career he had published studies of the behaviour of Tasmanian devils, and while in Tasmania he discussed with zoologists the concept of the carnivore guild and its functioning as a unit. His seminal thoughts and discussions contributed to the work and management directions which followed.

In his book (subtitled *An Analysis of Trends in Evolution, Adaptation and Behaviour*) Eisenberg first conceptualised the importance of studying a marsupial carnivore guild rather than individuals in isolation.[16] Menna Jones' resulting guild-structure findings derive from dentition studies carried out in the field and on skulls held in Australian collections. She showed that the relationship between devils and quolls evolved as one of direct competition. She sought to determine the role of such competition in structuring body size, habitat usage and diet. In general, species will space out in a habitat according to their own size and the size of their prey. It is called equal spacing.

A major finding was that, for this to be achieved, 'the spotted-tailed quoll had to redefine itself in an evolutionary sense'.[17] And it happened quickly: Jones puts the evolutionary timescale of

The spotted-tailed quoll competes with the devil for both live prey and carrion. Devils are frequently blamed for raiding poultry yards, when quolls are the more likely culprit. The spotted-tailed quoll, once commonly known as the tiger cat, is extinct across most of mainland Australia and no longer common in Tasmania. (Courtesy Dave Watts)

this at as little as 100 to 200 generations, a generation being two years. While devils and the small eastern quoll are sufficiently different in size as to have minimal dietary overlap, the larger spotted-tailed quoll is in the middle, and therefore in competition with both. Jones believes that this may explain why it is the rarest of the three. How did she arrive at these conclusions?

Skulls and skeletal material held in collections across Australia were measured. Particular attention was paid to dentition, with arrays of data compiled to create an index of tooth strength, as opposed to mere changes in tooth size over time. As well, the size and therefore strength of the *temporalis* muscles, the jaw-closing muscles, were measured by dimensions taken off the skulls. Tooth strength and jaw strength determine the size

This photograph offers rare proof of the predatory ability of the spotted-tailed quoll. This one has chased down and is killing a pademelon. (Courtesy Michael Good)

of prey a particular species can take. Analysis of the data indicated 'intense competition in Tasmania . . . anything other than equal spacing means two species are going to rub up against each other, hence enforced equal spacing. We got equal spacing in the jaw-closing muscle, tooth strength, and average prey size. That's pretty neat.'[18] Estimates are that the eastern quoll is three times as abundant and the (pre-disease) devil six times as abundant as the spotted-tailed quoll.

As the Australian continent dried out and heated up, the paucity of grazing or browsing vegetation shrank not only the megafauna but their replacements as well. Yet in this respect the Tasmanian devil is a veritable giant. Quolls aside, an adult male devil is up to 150 times larger than its closest marsupial relatives. They're also completely unalike, an indication of how varied the evolution of Australia's marsupial carnivores has been.

A male dusky antechinus in its favourite habitat of forest leaf litter. The tiny antechinus, weighing just 65 grams, is closely related to the Tasmanian devil. Males die within three weeks of mating, a feature of young devils since the onset of DFTD. (Courtesy W.E. Brown)

Thus the swamp antechinus (*Antechinus minimus*, 'smallest hedgehog equivalent') distributed across Tasmania and coastal Victoria, weighs about 65 grams. A strictly nocturnal insect eater and ground dweller—unlike the even smaller brown antechinus (*A. stuartii*) which likes to live in trees—this tiny marsupial is described as the smallest of the quolls. A biologically unusual feature of the antechinus is semelparousness, defined as reproducing or breeding only once in a lifetime. This is also a feature of the life cycles of animals including Pacific salmon, numerous spider species, squid and flying ants.

The little red antechinus (*Antechinus rosamondae*), which weighs about 40 grams and preys vigorously on lizards, seems to owe its precarious existence along the mid-north coast of Western

Australia to the fire-resistant, inedible woolly spinifex in which it lives.

The kowari (*Dasyuroides byrnei*) of central Australia, one of a number of marsupial species that become torpid during cold weather, is also a fierce hunter and vocally aggressive:

> A variety of sounds are produced, including an open-mouthed hissing and a loud, staccato chattering, both made in response to threats from predators or other kowaris . . . Vigorous tail-switching, reminiscent of an angry cat, is used as a threat display.[19]

The red-tailed phascogale (*Phascogale calura*) survives in Australia's arid centre through an impressive adaptation: it is immune to the poisonous plants upon which it feeds, as are the native carnivores which prey upon it. But the poison fluoroacetate, found in the Australian legumes *Gastrolobium* spp. and *Oxylobium* spp., kills introduced species.

An unusual adaptation is that of the long-tailed planigale (*Planigale ingrami*), one of the world's smallest mammals (the average male adult weighs 4.2 grams). Its head is flattened so that it can enter cracks and narrow spaces in search of the insects, lizards and small mammals which it attacks with ferocity.

A species as endangered as the spotted-tailed quoll, and matching the devil for size, now has full claim on being Tasmania's top order predator. The wedge-tailed eagle is a supreme hunter, one of the world's largest eagles. This majestic raptor, although distributed across Australia and New Guinea, is listed as vulnerable in Tasmania (subspecies *Aquila audax fleayi*). Its diet is a practical one, relying generally on possums, wallabies, rabbits, hares, birds and carrion.

The wedge-tailed eagle, one of the world's largest birds of prey, is considered to be Tasmania's top order predator now that the thylacine is extinct. Human alteration to the land, as well as direct persecution, have adversely affected both species for over two hundred years. The parent bird is on the right. (Courtesy W.E. Brown)

This means that wedge-tailed eagles and devils are direct competitors.

The eagle, like the thylacine, has long been demonised as a lamb-killer and has endured heavy persecution. Indeed, the formation in 1884 of the Buckland and Spring Bay Tiger and Eagle Extermination Society set in motion the Tasmanian parliamentary debates that were to signal the extermination of the thylacine. Well over a century later the 'wedgie' continues to be persecuted, with some rural Tasmanians taking a gun to it when they can.

4

'MADE FOR TRAVELLING ROUGH': DEVIL ECOLOGY

> Little Devil only wanted milk for several days given by a bottle while being cradled in my arms and loved being cuddled and would emit the most piercing sound when he had finished feeding. Little Devil lived in an old meat safe near the pot belly stove during the day in the kitchen and at night would come out for his meat and biscuits, but should anyone other than myself be in the kitchen he would retreat until all was quiet . . . Eventually the kitchen door was left ajar so he could come in and out at night, this happened for some time before he decided he was old enough to make it on his own. I guess like all teenagers he grew up. It will always remain one of my most treasured memories.
>
> DONNA COLEMAN, GORDON

In its evolutionary journey the Tasmanian devil has travelled remarkably well. And quickly: known devil fossils date back no more than 70 000 years and over that time the animal has undergone little change to its body plan other than dwarfism.

Its physical and behavioural characteristics helped ensure its success as one of the seven extant large-size marsupial carnivores of Australia and New Guinea.

The squat, muscular body and short strong legs enable it to lope long hours in search of food and, in the case of males, reproductive partners. Because they are large, the head and neck have increased functional significance in feeding. The devil's profuse, wiry vibrissae (whiskers) grow in patches from the tip of the chin to the back of the jawline, and are long enough to extend beyond shoulder width, acting as sensors during night movement, feeding and communication.

Devils, like dogs, have 42 teeth. (Cats have just 30.) Devils keep their original teeth, which continue to grow very slowly throughout the lifetime of the animals—they are not replaced.

The long claws are designed to dig efficiently, for denning and in search of food, and to firmly grip prey to facilitate chewing and gnawing. The sense of smell is acute and can detect food a good distance away.

This structural emphasis on feeding places the devil in the company of one of nature's iconic loners, the great white shark. A big old male devil has a shark-like forward torso, resulting in a great neck and head with a full but definite taper, providing immense power, out of proportion to the overall body.

The reproductive cycle of the devil is highly synchronised, but not inflexible. Female devils ovulate up to three times during the three-week breeding season, usually in late March, and copulation is almost continuous for up to five days at a time. The male goes to great lengths to keep other males away from his mating partner, keeping her prisoner in the copulation den with little chance to eat or drink. One thirsty male was observed dragging a female with him from a den to a water source and back to the den.[1] David Pemberton and Mount William ranger Steve Cronin once monitored a male and female in an underground mating den in the wild. The animals hadn't moved from it for eight days and nights, so, wondering if they had died, Pemberton dug a narrow hole through to the burrow and put his arm down, holding a small mirror. His colleague shone a torch onto the mirror, which revealed a threatening set of bared devil teeth.

However, the intensity of male competition generally ensures that during a normal breeding season males breed with more than one female. Menna Jones' studies of breeding indicate that females can be selective, and this means that in combination with multiple sperm donors the female optimises her chances of delivering the best available genetic offspring.

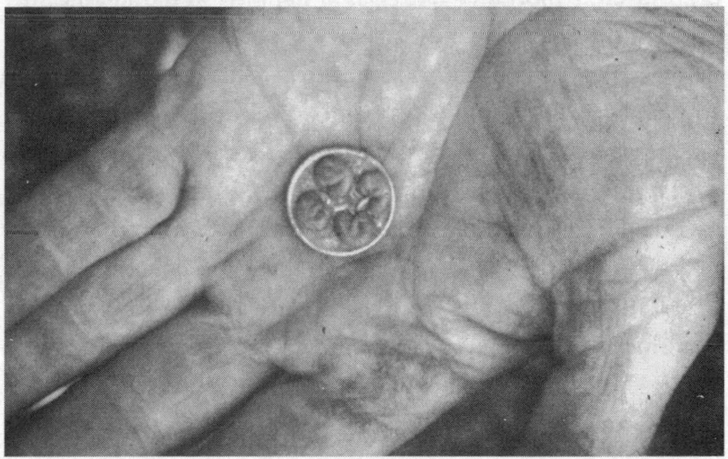

At birth, Tasmanian devils are tiny, as shown by these four newborns on a coin. (Courtesy Nick Mooney)

At birth a Tasmanian devil is no larger or heavier than a split pea. A prime four-year-old male is some 15 000 times heavier, at about 11 kilograms. (In comparison, from birth to maturity the average domestic cat increases in weight about 20-fold.) At six years the male will be dead, having sired perhaps sixteen offspring. An adult female, weighing about 7 kilograms, has on average four breeding seasons, producing about twelve offspring during her lifetime. The mating season is three weeks; pregnancy lasts just eighteen days; the young are dependent on their mother for at least nine months, which means the female then has little time to herself before the next mating season. This demanding cycle means devil populations can theoretically double in size each year, an excellent safeguard against high mortality in both juvenile and adult populations.

A female devil has four nipples in her marsupium and litters of three or four pups are common, which helps balance out the high juvenile mortality rate. David Pemberton found a healthy 80 per cent of two-year-olds carrying pouch young during his fieldwork study. The mother stands to give birth. Twenty or more tiny embryos leave the womb and travel up to the backward-opening pouch in a stream of mucus. The first arrivals clamp to the teats, which swell in their mouths, so that the newborns become firmly attached to their mother. This ensures they do not fall out of the pouch and is an important survival factor.

Devils are usually born in mid-April, that is, mid-autumn, ensuring they won't be weaned and have to face the world alone until long after Tasmania's challenging winter has passed. DFTD, however, has produced a dramatic shift in reproductive behaviour, with a scatter of births across the seasons rather than exclusively in autumn, and a high number of males competing

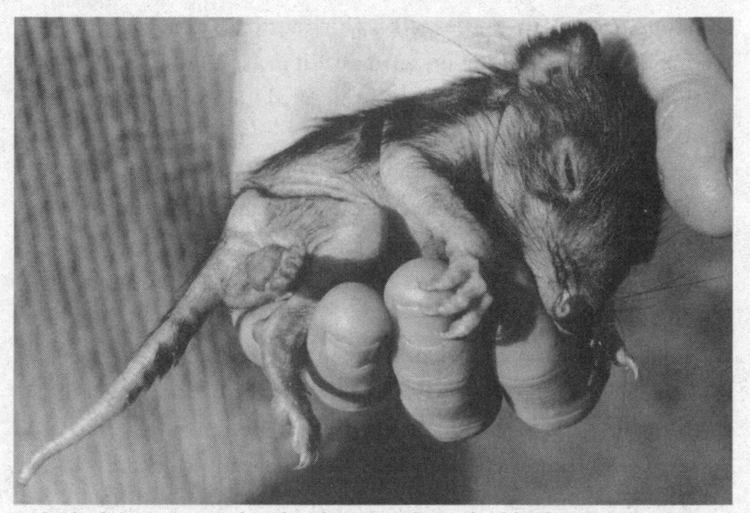

Baby devils begin to grow fur when they are twelve weeks old. Their sturdy tail is as long as their torso and has important functions including balance, storage of fat and communication. (Courtesy Collection Tasmanian Museum and Art Gallery)

for perhaps just one receptive female. This in turn increases the chance of spreading DFTD during copulation.

In 1934 David Fleay managed to breed devils in captivity and wrote a fine, precise account of it:

> In the first days of June four tiny, pink, naked and blind babies each a half inch in length had betaken themselves to their mother's pouch. Shortly after this the father was removed to bachelor quarters, for the mother now showed resentment at his presence by whining growls which rose abruptly in pitch and volume whenever the male attempted to enter the rock shelter. Early in August at the age of seven weeks the thick-set babies in the pouch had grown to a length of two and three quarter inches. They were still pink and hairless but now it could be seen that their tiny limbs moved actively as they

clung tenaciously to the teats within the pouch. They also made slight squeaking noises and with increasing bulk the hind quarters of one quadruplet projected from the pouch as the mother moved about.

Meanwhile she had become somewhat fastidious for a Devil, disdaining raw meat but delighting in rats, birds, eggs, frogs and rabbit heads. Towards the middle of August a great change came over the appearance of the youngsters as the ear tips and then other regions of the skin began to show dark pigment. The pouch too, developing with the family, was far more relaxed and roomy. At eleven weeks the dark pigment of the young had become sufficiently pronounced to throw into strong contrast the future white chest and rump markings. The quiet nervous mother accepted the frequent handling with no sign of resentment. Progress of the little Devils was now quite phenomenal and on October 1st at fifteen weeks of age they first released their till then continuous grip on the teats. They were well furred and their eyes had opened. From these observations it is obvious that the mother must carry her cumbersome family with her for at least fifteen weeks after their birth. But from this time on the youngsters may be left at home in the nest, allowing the mother the freedom necessary for successful 'scrounging'.

When lifted away from the parent the youngsters uttered anxious yapping cries and on being released again clung quickly to the fur of her sides with teeth and fingers—the fore-paws having unusual grasping powers so that the young Devils are expert climbers. On being disturbed from sleep when sheltered by the mother's body, the little fellows lost no time in gripping her extended teats, from which it was almost impossible to dislodge them until firm pressure with a finger-tip over the nostrils caused their mouths to open. At the age

> of eighteen weeks the 'play age' was apparent . . . At twenty weeks they were seven and three quarter inches in body length with small tails adding a further three inches. They still clung tenaciously to the mother's teats when drinking . . . It was five months before they ceased to rely on their mother's milk for nourishment and unfortunately we lost two of them before they had abandoned the maternal apron strings. One squeezed through the chain netting of the enclosure and was never heard of again while the other sickened and died. The mother and remaining two youngsters showed the thorough scavenging traits of their kind by immediately devouring the whole carcass, except the head, of their deceased relative even though food was plentiful.[2]

Devils are weaned in summer, between December and February, after which they disperse widely, but with a higher proportion of females remaining in the natal areas. Up to 60 per cent die before reaching maturity, according to Guiler. Even so, the sudden increase in numbers over summer can give the appearance of a plague, because juveniles are more crepuscularly active than adults and at dusk are regularly seen on roads, scavenging in paddocks or on beaches and around farm complexes.

Newly weaned devils become solitary, although there is some evidence of initial communal travelling. They are agile foragers, taking a wide variety of small invertebrates and vertebrates, and their excellent climbing ability enables them to obtain food from trees, such as grubs, eggs from birds' nests and possums. They are fully grown and mature by the age of two.

The uniqueness of a solitary animal surviving through communal feeding sets the devil apart from other carnivores. Young devils quickly learn to congregate at the site of a carcass,

Young devils are agile foragers and good climbers. This picture was taken at Mrs Roberts' Beaumaris Zoo in the early 1900s. (Courtesy Collection Tasmanian Museum and Art Gallery)

drawn by the scent and, just as importantly, the vocalisations of those that have already arrived at the site. Conflict over a carcass is avoided through a ritualised behaviour ensemble.

Young devils, because they are active at dusk, have the advantage of arriving at food sources before the more competitive adults. This, however, puts them into competition with spotted-tailed quolls, which are also active diurnal feeders.

Feeding devils communicate with each other through a range of visual postures, vocalisations and a suite of chemical signals.

It was once assumed that dasyurids made little use of visual communications because they are nocturnal. Devils have night-adapted eyesight and their white chest and rump flashes are distinctly visible at night. Interaction between devils at feeding sites takes the form of a ritualised contest, with the dominant feeder not being displaced until it has gorged itself.

A devil eats up to 40 per cent of its body weight per meal every two to three days. Eating such a large quantity of food in a short space of time—about half an hour on average—often results in the animal waddling off with a distended belly and lying down not far from the feeding site; a devil in this state is easy to approach. It is likely that the absence of other large predators has facilitated this form of feeding, even during the long period when thylacines were Australia's largest carnivore. This lends some support to the belief that thylacines ate only choice parts of their prey, leaving the rest to devils and other scavengers.

Being a scavenger able to digest a wide variety of food matter—flesh, fish, bone, invertebrates, fruit, vegetation—was an advantage to the devil's survival. It may also be that the thylacine's narrower food base (and less productive breeding cycle) meant that it existed in comparatively lower numbers, leaving it more vulnerable to changed circumstances (human predation) than the devil.

The number of devils feeding together is generally determined by the size of a carcass: groups of two to five are common. The first arrival is the dominant feeder (unlike communal hyaena feeding, where the higher ranking clan members feed before subordinates), which makes way for a challenger once it has gorged itself. The size of the carcass affects the extent to which the feeding devil will chase off a challenger: the feeder defends the amount of food it needs, not the entire carcass. Satiation, rather than dominance, is the most likely information conveyed by the ritualised interactions. This means all devils, small and large, resident and transient, can feed together. It is an efficient way of sustaining a population.

David Pemberton was the first to make scientific field studies of devils feeding in the wild. The absence of unrestrained aggression while feeding, and the complex behaviour occurring in its place, was a critical discovery, overturning popular (and some professional) perceptions of the animal, as in this supposedly informed 1984 account of feeding devils by a popular natural history author: 'They behave like a brawling mob, having, so far as anyone knows, virtually no social organisation or restraining instincts'.[3]

Pemberton recorded only one instance of physical injury in 119 interactions during feeding, with one animal chasing and biting another on the rump. On two occasions he also observed jaw-wrestling, where devils stood on their hind legs with forepaws on each other's shoulders or chest and their jaws interlocked. The animals vocalised constantly while shaking their heads from side to side. Although there was no obvious physical damage to the animals, the nature of the interaction appeared as if it could have caused extensive damage to muzzles or jaws. On each occasion the defeated animal ran off into the bush with its tail in the air and fur fully erect, with the winner pursuing it and biting its rump whenever it was close enough.[4]

Examination of 150 trapped animals showed that 6 per cent had suffered injuries consistent with fighting during feeding or breeding, showing enough damage to the flesh of the face to leave teeth visible. These, however, were aged males with lame hindquarters and extensive hair loss on rump and tail, suggesting a physical deterioration other than that caused by intra-specific aggression. A third showed some form of wound, such as puncture holes on the back and rump and shortened, hairless tails. It doesn't automatically mean that all the wounds were obtained while feeding; in fact, most wounds occur during the breeding season.

Pemberton recorded eight vocalisations between initiators and recipients at carcasses: a 'snort' made by expelling air through nostrils and mouth; a short, deep, low-intensity 'humf growl', often repeated; a short, deep, high-energy 'bark', seldom repeated; a 'clap' made by snapping the jaws together; monotone, vibrato or crescendo 'growl-whines'; a 'screech', associated with defeat; a 'sneeze' and a 'yip'.[5]

Pemberton also identified some 20 postures that reinforce the elaborate nature of the feeding interactions. They include:

- Neck threat—one nips at the neck area of the other, without making contact. These nips are repeated and the recipient responds by shouldering the initiator or attacking it face on.
- Gape—animals open their jaws for a few seconds as inter-actions take place.
- Lying down—the initiator of the behaviour lies down on its belly with the fore and hind feet extended. These animals are in full view of the possessor of the carcass, but do not physically interact with it.
- Sitting—the initiator sits and stares in the direction of the recipient, often combining this with gaping and lying down.
- Head and tail positions—these vary from a frequent 'head-up-tail-down' posture to a less frequent 'head-up-tail-straight'

Devils regularly use communal latrines, as this area of scats clearly shows. The latrines are believed to function like community noticeboards. (Courtesy Nick Mooney)

posture. Their intensity seems to be reflected by the degree to which the fur on the tail is raised. In some circumstances the legs are bent sufficiently for the belly to scrape on the ground.

- Tripod—the initiator raises one front paw off the ground while facing another animal, with the head held up and the tail down.
- Stiff legged—the initiator moves forward a short distance without flexing the joints of the legs.
- Urinating—the initiator urinates in full view of the recipient, sometimes in conjunction with stiff legs or gape. This action may also facilitate a chemical signal to the recipient.
- Ano-genital drag—the animal presses its anus and genitals against the ground and drags itself along with its forepaws, while holding its chest off the ground. This posture is

sometimes accompanied by gaping and an erect tail. It may also be associated with the transmission of chemical signals. Characteristic drag lines are often seen at communal latrine sites (and are easily mistaken for wallaby tail marks).

Devil literature sometimes refers to ano-genital dragging as cloacal dragging, but this term is inaccurate. Males as well as females engage in dragging, and they begin doing so well before being weaned. Although scent marking is associated with territoriality in many animal species, the function of dragging is not fully known.

Dens and home ranges are key aspects to understanding devil behaviour in the wild. Devils regularly use three or four dens, preferring wombat burrows, dense vegetation near creeks, thick grass tussocks, and caves where they are available. Old wombat burrows and caves are favoured maternity dens, no doubt because of their relative security from predators. Once established in dens adults tend to use them for life.

Devil pups cling to their mother when she needs to transport them. (David Fleay)

While adult devils alternate between dens, dependent denned young are not generally moved from one to another. Thus, while mothers with young have a fixed location, all other devils move around. It is apparent from this that the locality of dens forms an integral aspect of the spatial organisation of devils, as well as being critical to survival of the young. In this way the devil is more dependent on its den than its larder. The same applies to the wolverine, which also dens its young in a fixed locality. Studies have shown that habitat disruption, while it may not negatively affect the local food supply, exposes these natal dens and increases mortality as a result. When this occurs, the mother moves with the young clinging to her back; this greatly increases their vulnerability.

Habitat interference affects animals by altering the refuges where they breed, raise young and rest. For the devil this could be critical. Maternity dens are carefully selected to provide a safe haven from the elements and from scavengers. Young devils get cold easily and need the warmth of their nests and the sun. Favoured dens are strongly protected and may have existed for centuries. Destroying them through, for example, land clearance, disrupts population stability. Reduction in suitable denning habitat has significantly affected the wolverine.

The location and shape of home ranges appear to be controlled by the distribution of food, primarily wallabies and pademelons, both of which are present in large numbers in many parts of Tasmania. Both male and female adult devils are active between sunset and sunrise, travelling up to 16 kilometres a night in search of food. In this way they cover the extent of the home range, generally in a circular pattern. Males and females have similar sized home ranges, which is unusual in sexually dimorphic, solitary carnivores. Larger males obtain the additional food they need by eating for longer. While engaged on his doctoral research David Pemberton charted one devil's home range, which 'took in three hectares of the Musselroe Bay holiday village and tip site'.[6]

In the 10 000 years in which they have lived isolated in Tasmania, the devil's predator competitors—thylacine, wedge-tailed eagle, spotted-tailed quoll—have not threatened the survival of the species. Furthermore, apart from during the mating season and occasionally while feeding, devils do not engage in combat with one another. This is why, even though solitary, they are able to live together in high densities and share overlapping home ranges. Thirty devils weighing a combined 250 kilograms equates to many fewer bears or hyaenas occupying the same space. Even so, consider a Tasmanian farmer's anxiety if he had six hyaenas living in his back paddocks.

5

DEVILS AND EUROPEANS, 1803–1933

The devil stories I was told as a farm boy were shot through with menace. Devils were as dangerous as the name suggested. In large packs they stalked and harassed pregnant, sick or dying cows and horses. If the stricken animal refused to fall the pack leaders would bite clean through its front legs and the rest would swoop in . . . I can see now that the old men and women who said they knew for a fact that devils attacked animals many times their size were just trying to protect me from the dangers of the bush. But they were also voicing the fears they had inherited from those early convicts who were being literal when they named the carnivores they had newly encountered after the Prince of Darkness.

RODNEY CROOME, HOBART

In 1803 an attempt to establish a convict colony at Port Phillip Bay (subsequently Melbourne) proved unsuccessful and led instead to the settlement of the penal colony of Van Diemen's Land. A member of the small founding group was George

Prideaux Harris who, in Hobart Town, worked as a magistrate, journalist, surveyor and natural historian. He became the first European to describe and classify the devil, naming the squat, peculiar little animal *Didelphis ursina*. The name he gave the genus echoes the American opossum, while the species name was intended to reflect its bearlike (ursine) qualities, not least the small round ears. Harris correctly noted a number of similarities between the devil and the thylacine, one being a marsupial trademark, that the rear heels of both are long and callous.

George Prideaux Harris, after whom the devil is named, wrote the first description of the animal in 1806:

> These animals were very common on our first settling at Hobart Town, and were particularly destructive to poultry, &c. They, however, furnished the convicts with a fresh meal, and the taste was said to be not unlike veal. As the settlement increased, and the ground became cleared, they were driven from their haunts near the town to the deeper recesses of forests yet explored. They are, however, easily procured by setting a trap in the most unfrequented parts of the woods, baited with raw flesh, all kinds of which they eat indiscriminately and voraciously; they also, it is probable, prey on dead fish, blubber, &c. as their tracks are frequently found on the sands of the sea shore.
>
> In a state of confinement, they appear to be untameably [sic] savage; biting severely, and uttering at the same time a low yelling growl. A male and female, which I kept for a couple of months chained together in an empty cask, were continually fighting; their quarrels began as soon as it was dark (as they slept all day), and continued throughout the night almost without intermission, accompanied with a kind of hollow barking, not unlike a dog, and sometimes a sudden

George Prideaux Harris, the deputy Surveyor General in the first party of Britons to settle Van Diemen's Land, drew these devil and thylacine sketches for the Linnean Society of London in 1806. (Courtesy Linnean Society of London)

kind of snorting, as if the breath was retained a considerable time, and then suddenly expelled. The female generally conquered. They frequently sat on their hind parts, and used their fore paws to convey food to their mouths. The muscles of their jaws were very strong, as they cracked the largest bones with ease asunder; and many of their actions, as well as

> their gait, strikingly resembled those of the bear . . . Its vulgar name is the Native Devil.[1]

Fifty years after Harris wrote his description of the devil, the English naturalist and artist John Gould compiled his three-volume *Mammals of Australia* and his seven-volume *Birds of Australia*, both brilliant and enduring records. Gould, who predicted the thylacine's demise nearly a century before its extinction, wrote of the Tasmanian devil:

> [I]ts black colouring and unsightly appearance obtained for it the trivial names of Devil and Native Devil. It has now become so scarce in all the cultivated districts, that it is rarely if ever, seen there in a state of nature; there are yet, however, large districts in Van Diemen's Land untrodden by man; and such

Famous naturalist John Gould described many species of Australian fauna in the mid-nineteenth century and had an array of artists working for him, including Henry Richter who drew this devil from a live specimen in London's Zoological Society menagerie. Unusually for a Gould work of art, it's anatomically incorrect, having a small, underslung jaw. (Gould, *Mammals of Australia*)

> localities, particularly the rocky gullies and vast forests on the western side of the island, afford it a secure retreat. During my visit to the continent of Australia I met with no evidence that the animal is to be found in any of its colonies, consequently Tasmania alone must be regarded as its native habitat.
>
> In its disposition it is untameable and savage in the extreme, and is not only destructive to the smaller kangaroos and other native quadrupeds, but assails the sheep-folds and hen-roosts whenever an opportunity occurs for its entering upon its destructive errand.
>
> Although the animal has been well known for so many years, little or nothing more has been recorded respecting it than that which appeared in the ninth volume of the Linnean Society's Transaction from the pen of Mr Harris . . .[2]

Writing in 1880, author and artist Louisa Anne Meredith did much to publicise the new British colony. Her books were very popular in England and her chatty, hotchpotch style says as much about Victorian readers as it does the animal under scrutiny:

> [I]f anyone desires to see a blacker, uglier, more savage, and more untameable beast than our 'Devil', he must be difficult to please—that's my opinion. I suppose those who bestowed such a name on him had pretty good reasons for it, and knew that they only gave the devil his due . . . I've heard people say in joke, of others who had very wide mouths, that, when they gaped, their heads were off; but it seems true of this animal, his jaws open to such an extent, and a murderous set of fangs they show when they do open!
>
> The head, which is flat, broad, very ugly, and with little skull-room for brains, takes up one-third the whole length of the beast, which is usually from a foot and a-half to two feet, some being larger. The tail sticks stiffly out, as if made of

wood, the feet are something like a dog's, only more sprawly, and with very big claws. It is an awkward beast, and cannot go much of a pace at the fastest. On fairish ground, a man can easily run one down.

One day I was out with Papa in the back-run, and we found a devil. I started full tilt after him, and came two or three good croppers amongst the rocks to begin with, but I held on, till all of a sudden he stopped short—I couldn't, so I jumped right over him. He gave a vicious snap at my legs with his big jaws, but, luckily for me, he was a second too late. I turned and knocked him over, and papa came up and finished him—finished killing him, I mean. We don't show the brutes any mercy; they do too much mischief. The young lambs stand no chance at all with them. So we hunt them down, or set traps, or dig pitfalls—any and every way we can destroy them we do. Why, one winter, some years ago, one of Papa's shepherds caught nearly one hundred and fifty! They seem to go about in families or parties; for when you catch one, you are tolerably certain of getting six or seven more, one after another, and then perhaps you will not hear of any for a good while. Of course they are much scarcer than formerly, and a very lucky thing, too.

I don't think I mentioned the fur—but it is not fur, it's longish, very coarse, black hair, almost like horse-hair; and then as to fleas, *they swarm!* One of the men brought a dead one to the house one day for Mamma, and it was laid in the garden. Mamma and Lina were soon down on their knees beside it, peeping at its eyes and teeth and ears and all the rest of it; when Lina said, 'Oh, look; how very curious! There are small, brown scales, like a coat of mail, all over it, under the hair'. Mamma looked where Lina had parted the long hair, and *didn't* she jump! Lina's coat of mail was just a coat of fleas.

> The *post-mortem* examination was cut very short, I assure you, the 'subject' summarily disposed of, and two or three buckets of water poured on the place where it had lain. A pleasant kind of thing for a pet!
>
> There are two sorts of devils—one is all black, the other has a white tail-tip and a white mark like a cross down the throat and between the fore-legs; but one is just as hideous as the other. I believe you cannot tame them, and I am very sure I shall never try. People who have made the attempt say they are as stupid as they are ferocious, and never seem to know one person more than another, but growl and bite at all alike.[3]

For all her apparently direct association with devils, many of her descriptions are clearly inaccurate, and it's of interest that the drawing of the animal accompanying Meredith's text is a freehand copy of Gould's original, unattributed and minus the background devils. She—or another—likewise copied the famous Gould lithograph of a thylacine pair, an animal admittedly much harder to locate, let alone sketch.

Louisa Anne Meredith died in 1895. In that year, wealthy Hobart socialite Mary Roberts opened a private zoo at Beaumaris House, close to the town. The contrast between the women is stark. Not only did Mary Roberts like devils, she bred them, and in doing so helped shift its image from diabolical and satanic to merely animal. Roberts achieved international fame for her devotion to animal causes, through activities such as her founding of the Anti-Plumage League. But she also had a highly developed business sense, importing wildlife from all over the world, while exporting whatever Tasmanian fauna she could.

Not unlike George Harris almost a century earlier, in 1915 Mary Roberts wrote about the devil for an academic British audience, this time for the London Zoological Society. It's an important and accurate document, given the lack of written information about the devil then and universal ignorance of it. The article is titled 'The Keeping and Breeding of Tasmanian Devils':

> Until I was asked by Mr. A. S. Le Souëf, Director of the Zoological Gardens, Moore Park, New South Wales, early in 1910 to obtain, if possible, Tasmanian Tigers (*Thylacinus cynocephalus*) and Devils (*Sarcophilus harrisii*) for the London Zoological Society, I had never thought of keeping either of these animals in my collection; in fact, they were quite unknown to me except as museum specimens, although I had frequently visited remote parts of our island. I have vivid recollections, however, of how, when a young girl at boarding-school in the late [eighteen] forties, some of the girls from Bothwell, near the Lake District, used to give graphic and terrifying accounts of the Tasmanian Devils with their double row of teeth. This belief is not yet exploded, as it was impressed upon me lately with the utmost confidence by a country visitor that such was the case; he not only believed, but said 'he had seen'. The teeth have been described to me by a scientist as truncated.
>
> Shortly after hearing from Mr. Le Souëf, by means of advertising, writing, etc. I obtained three for the London Society, and having then become thoroughly interested I determined to keep some myself. Since that time a large number have passed through my hands, and more than once I have been 'a woman possessed of seven devils'.
>
> In April 1911 I received a family (a mother and four young), and again in September of the same year a similar lot arrived. The former were very young, and I had the

Mary Roberts owned and operated Beaumaris Zoo in Hobart between 1895 and her death in 1921. She collected animals and birds from all over the world and also maintained a thriving business exporting Australian fauna. She wrote that Tasmanian devils were her favourite creatures. Her insights into their behaviour were in marked contrast to public perceptions of them as stinking vermin. (Courtesy Collection Tasmanian Museum and Art Gallery)

opportunity of watching their growth almost from their first appearance when partly protruding from the pouch. When sending them, the trapper wrote that 'the mother was so quiet, I need not be afraid to pick her up in my arms'. The little ones hung from her pouch (heads hidden in it), and she lay still and motionless as if afraid of hurting them by moving, and allowed me to stroke her head with my hand. However timid they may be, and undoubtedly they are extremely so, growling and showing their teeth when frightened, they always evince this gentleness and stillness when nursing little ones.

The skin of the young, on arrival, had the appearance of a slate-coloured kid glove, the tail darker towards the tip. The hair could be seen growing black and velvety from the head downwards, the latter being hidden in the pouch for some days, and it was interesting to note the progress of the growth of the hair from day to day. The shoulders were covered while the hindquarters were almost, or quite, bare, although a faint streak of white was discernible where the white markings were to come later on. At this early stage, should the mother get up to move about, which she rarely does in the daytime, the young somehow scramble into the pouch again.

This family went later to the London Society, but the second, which came on the 16th of September, I kept for my own pleasure, with the exception of the mother; as she had lost a foot when being trapped, I thought it best to have her destroyed later on. Unfortunately, when they were about half grown one escaped into the garden, and the next morning her mutilated remains were found—she had fallen a victim to our two fox-terriers. The three survivors have been ever since an unfailing source of interest and amusement to my family, to visitors, and myself. When a bone or piece of meat was thrown to them a tug-of-war was always the result, and sometimes a chase into one door and out of the other of the little cave. At other times, while one has been holding on to a bone held in my hand, I have lifted it completely off the ground, while another would cling on round the waist and try to pull it down.

Many visitors from the Commonwealth have heard such exaggerated accounts of the ferocity and ugliness of the Tasmanian Devil (others, again, have believed it to be a myth), that they sometimes express surprise when they see them so lively, sprightly and excited, running out to my call; they then remark, 'the devil is not so black as he is painted'.

Two of these Devils were latterly kept together as a pair, and for the purposes of this article I will call them Billy and

Tasmanian devils at Beaumaris Zoo, c. 1910. (Courtesy Collection Tasmanian Museum and Art Gallery)

Truganini, after the last two survivors of our lost Tasmanian race.[4] These showed no disposition to breed until April 1913, and my observation of them and of many others that I have had in my keeping is, that the disinclination to take up maternal duties is always on the part of the female. I then noticed suddenly a decided change—that Billy would not allow her to come out of their little den; if she did venture when called to be fed, or at other times, he immediately attacked her and would drag her back by the ear, or any other part, but although otherwise cruel, he would carry food in to her. When I called her, it was pitiable to hear her whining; but it was of no avail, for Billy was a relentless tyrant and kept her in strict seclusion for quite ten or twelve days; then early in May he allowed her to be free once more. From thence onward, although they were sometimes peaceable and affectionate, the balance of power was completely on

Truganini's side; she constantly resented his approach by biting and snarling at him: it seemed as if coming events cast their shadows before, and she instinctively felt that he would do the young some injury. From now her pouch was anxiously scanned day by day, but it was some time before I could be sure that it was gradually enlarging. I had been advised by Dr Hornaday, of the New York Zoological Park, that if ever the Tigers or Devils were likely to have young, to remove the male, and as soon as I was certain, I had Billy taken away and placed with the other member of the family. This made Truganini most unhappy, as he was near enough for her to hear him, besides which, the two males fought; so, being cautioned by my family that perhaps my interference might cause a disaster, I yielded and replaced him, doing so with many misgivings. Matters went on much the same until late in September, when to my delight a tail, and at other times part of a small body, could be seen sticking out of the pouch, more especially when she sat up to wash her face, or rolled upon her back; unlike domestic cats, the devils use both paws for washing, placing them together and thus making a cup-like depression which, when thoroughly licked, is rubbed well over the face. Everything looked very promising on the Sunday before Michaelmas Day, when I noticed Truganini carrying large bunches of straw about in her mouth, evidently seeking for a retired place to make a bed, and we had already placed some fern logs in a corner of their yard. As Billy would follow her about and interfere, I had a box put down with a hole cut in the side that she might hide under; but it was of no use, as where she went he would also go, and a scrimmage was the inevitable result. Early next morning, with many misgivings I left home for ten days, only to find on my return that her pouch was empty and that the young had disappeared, and as no remains whatever had been found, I could only conclude that they had been eaten by Billy.

Thus ended all my hopes and anticipations for 1913. I have not so far related an incident that took place just before the breeding-season. Being hopeful that Truganini might have young in her pouch, and my assistant being as usual very busy, Professor T. T. Flynn, of the Tasmanian University, who is always interested in our marsupials, kindly offered to examine her pouch. As soon as an attempt was made to catch her, Billy grasped the position of affairs and fought to defend her with all his might, even getting behind her in the little cave, putting a paw on each shoulder and holding her tightly, lest she might get into what appeared to him to be the danger zone. By dint of perseverance and a little strategy he was outwitted at last, but our hopes were doomed to disappointment.

Truganini has now passed through another period of retirement, and I am hoping to record shortly a greater measure of success for 1914.

I cannot close this article without a few words in defence of the Tasmanian Devil, as I am sure that it is more or less 'misunderstood', and the article with photograph published in the 'Royal Magazine' for October 1913 under the name of L. R. Brightwell, F. Z. S., is, I consider, greatly exaggerated both as regards their appearance and character, viz., 'They are well named, for they tear everything, even sheep, to pieces if they get the chance'.

On several occasions when one of mine has escaped, the only mischief done has been the destruction of a fowl or a duck or two. It would have been just as easy for a wallaby to have been killed if they had had the inclination, about which our fox-terriers would not have hesitated for a minute if a chance had occurred. When in transit to London last year one

> escaped, and I have been told by the chief officer of the vessel that 'the passengers were much alarmed as there were children on board, and someone went about with a revolver'. Later I came across the butcher who was in charge at the time, and he appeared to have been rather amused than otherwise, and told me the missing one was discovered at last sleeping under the berth of one of the sailors! I don't wonder, with the reputation that the devils have, that the passengers were alarmed.[5]

Mary Roberts had more luck in 1914 when Truganini gave birth to three babies. Billy was again the father but was now kept away from the maternal enclosure. Roberts compiled diary notes, as with these examples:

> 29th—All three playing like puppies, biting each other and pulling one another about by the ears . . .
>
> 30th—Whole family hanging from the mother as she ran out, and one hardly knows which to admire most, her patience and endurance, or the hardihood of the young in holding on and submitting to so much knocking about. The whole process seems very casual and most remarkable . . . The baby devils had the sense of smell very strongly developed; immediately I approached, their nostrils would begin to work and a vigorous sniffing would go on. They were also expert climbers, and although I had some specially constructed yards made, they would get up the wire-netting and walk along the top rail quite easily; at other times they would climb a pear-tree growing in their enclosure and sit in the branches like cats.[6]

Her article concludes with a section headed 'General Remarks':

> I have always found devils rather fond of a bath; quite recently, going down to their yard after an illness and finding only a

drinking vessel, I ordered a larger one to be put in, and they showed their pleasure by going in at once, sometimes two at a time. I have occasionally poured water from a can over them, when they would run to and fro under it with much enjoyment.

Their sight in daylight is rather defective; they seem to pick up their food more readily by smelling than by seeing, and I think they can see objects better at a distance.

At the present time I have six running together, my own three and three that I bought when in their mother's pouch. All are tame, frolicsome, and lively. I can go in and have a bit of fun with them, and when I am outside their enclosure they frequently climb the wire-netting to the height of nearly six feet, and get their little black faces close to mine with evident delight. We have tried more than once to get them photographed, but it is impossible to keep them quiet, they are on for a scamper all the time. Recently an adult escaped, and it was discovered by a passing school-boy sitting on a high fence bordering the street, under the shade of some elm-trees, many people passing on the foot-path without observing it. They are, however, always very timid when coming down.

They are fond of the sun, and look well when basking in it, the rays shining through make their ears appear a bright red, fore-feet parallel with the head, hind-quarters quite flat on the ground and turned out at right angles, somewhat as a frog.

My sympathy with my little black 'brothers and sisters' is intense, probably evoked by having suffered much mentally owing to the gross cruelties which have come under my notice, the result of capturing them in traps. Frequently three or four have been sent to me in a crate, only to find later on one with a foot shot off or a broken leg. In a consignment received some time ago, a dead one was found; it bore unmistakable signs of

> a snare previously, round the neck, one foot was gone (an old injury), and finally a recently smashed leg much swollen, the cause of death. I communicated with the S.P.C.A., and since then have had none from that district.
>
> I have derived much pleasure from studying the habits and disposition of the Tasmanian Devils, and have found that they respond to kindness, and certainly show affection and pleasure when I approach them. I have been led to believe that no case of their breeding in captivity has been recorded, and certainly not in Tasmania.
>
> Others who do not know or understand them may think of them as they like, but I, who love them, and have had considerable experience in keeping most of our marsupials, from the Thylacine down to the Opossum Mouse (*Dromica nana*), will always regard them as first favourites, my little black playmates.[7]

Mary Roberts wasn't a trained scientist. But her Beaumaris Zoo not only popularised native animals until then considered loathsome, dangerous and expendable; it also attracted those few scientists who had begun devoting their energies to understanding and protecting the island's fauna. One was Clive Lord, Director of the Tasmanian Museum, who in 1918 compiled a list of about 50 known descriptions, classifications and drawings of the devil. He expressed concern that native species such as the devil were decreasing in numbers while very little was known about them.

Another was Professor T. T. (Theodore Thomson) Flynn, who occupies an important place in Australian zoology as a pioneering early twentieth-century mammalogist. His works on the embryology and early development of native animals are

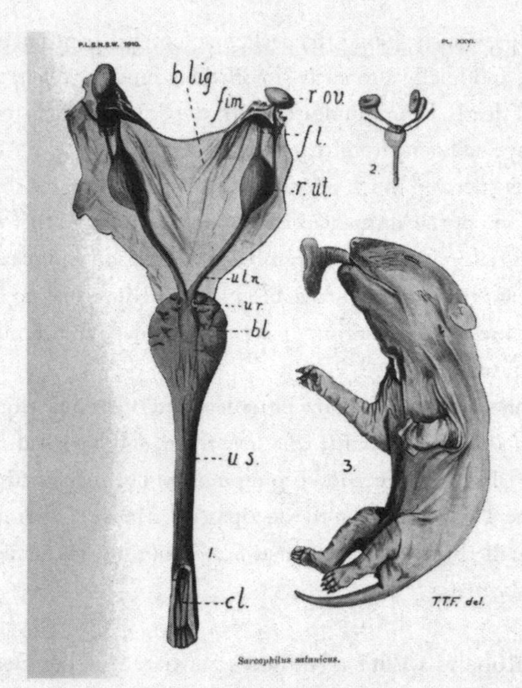

Professor T.T. (Theodore) Flynn, a biology lecturer and researcher at the University of Tasmania from 1909 to 1930, was the father of actor Errol Flynn. Theodore undertook pioneering laboratory work on devils, one result being this fine natural history illustration of the urogenital system and pup on teat. (Courtesy Collection Tasmanian Museum and Art Gallery)

rightly described as classics. In 1909 he had become the inaugural Professor of Biology at the University of Tasmania and for 20 years remained devoted to his research. He was originally offered his academic position in Tasmania to study marsupial diseases but, not finding any, he branched out into other aspects of zoology, including keeping devils in his backyard.

One of Flynn's early publications was 'Contributions to a Knowledge of the Anatomy and Development of the Marsupiala:

No. I. The Genitalia of *Sarcophilus satanicus*'. His research derived from a single female devil, the first ever to come into his possession—not from Beaumaris Zoo but from Clive Lord. In his introduction Flynn noted that his intention had been to study a number of specimens before publishing his results, but their 'increased scarcity' decided him otherwise.[8]

The research itself was obviously not easy, Flynn noting the 'unfortunate lack of original communications and papers in Tasmania'.[9] His introductory notes are illuminating:

> The specimen of *Sarcophilus satanicus,* of whose genital organs this communication is a description, was forwarded to me through the kind offices of Mr J. E. C. Lord . . . This is the only female which I have as yet obtained and I had originally intended that its description should wait until further specimens had come to hand; the increased scarcity, however, of these animals, together with the discovery of a number of interesting and significant points in the morphology of the genital organs, has influenced me to publish the results earlier than otherwise would have been the case. Portions of the paper can as yet be regarded only as preliminary notes. This is due, in the first place, to scarcity of material, and, in the second, to an unfortunate lack of original communications and papers in Tasmania.

His short, preparatory description further reveals the difficulties of conducting pioneering scientific work under the conditions he experienced:

> The specimen was a full-grown female, with three fairly advanced young in the pouch. All had been dead for two days. The pouch-young were fixed entire in corrosive-sublimate-acetic-solution, the genital organs of the mother in picro-sulphuric solution. In this latter case, on sectioning, it was

> found that what blood there was in the vessels had hardened so much, that it was only with extreme care and difficulty that sections could be cut at all. The hopeless gapping of the razor-edge, with consequent damage to the sections, is well indicated in Fig. 10.[10]

Guiler provides an interesting snapshot of Flynn and Roberts:

> Flynn was a very powerful personality and full of drive and energy which led him into many adventures, creditable and otherwise. He was often at the Zoo in his early days in Tasmania but the [Roberts] Diary entries show a declining enthusiasm and in August 1918 Mrs Roberts records that she had sent an account to Prof. Flynn for the devils, adding that she expected to be paid. In October 1918 she records that she rang the University 're the skeletons Flynn has'; he was to ring back the next day but did not do so. Possibly the non-payment for specimens may have been the cause of Mrs Roberts' annoyance with Flynn but it is not surprising that there was a cooling in their friendship, as Mrs Roberts was most fastidious in all her dealings and Flynn most casual and unbusiness-like in his.[11]

Among Flynn's 'adventures' were rumoured clandestine sales of thylacines. His family life was messy, including estrangement from his wife and considerable difficulty in managing his head-strong, wild, sexually charged son Errol, destined to become the dashing star of more than 50 Hollywood movies. The first chapter of Errol Flynn's autobiography is called 'Tasmanian Devil, 1909–1927', and it was his Hollywood studio, Warner Bros., that created the irrepressible *Looney Tunes* cartoon character, Taz the Tasmanian Devil.

Tasmanian Museum director Clive Lord, in his own writings on the devil, observed:

> Its hardy nature both in captivity and in its wild state cause one to wonder how it came about that this species became extinct on the mainland within comparatively recent times . . . In the rougher sections [of Tasmania] this species exists in fair numbers and there is every prospect of it remaining an inhabitant of such places for years to come.[12]

Lord also sounded what might be called an optimistic warning, one that has still not been resolved 200 years after George Prideaux Harris wrote his description of the devil. Towards the end of his life Lord wrote: 'We, as Australians, have been placed in charge of a wonderful heritage, and it rests with us to respond to the trusteeship which has been granted us.'[13]

Clive Lord died in 1933 and so did official interest in the devil.

6

DEMONISING THE DEVIL

Tasmanian devils have been the subject of an extraordinarily intense and innovative research and management program that has seen the ravages of DFTD understood and tempered, and insurance populations established behind fences and on islands. The immediate future of the Tasmanian devil thus seems assured. But, what of the longer term?

Saving the Tasmanian Devil: Recovery Through Science-Based Management, CSIRO Publishing, 2019

Multiple perceptions of, and attitudes towards, the Tasmanian devil have been shown in previous chapters. This new chapter and the two following expand considerably on the non-scientific historical relationship between this singular species and people, through media not readily obtainable when the first editions of the book were published. As will be demonstrated, despite relentless demonising, persecution and exploitation for some two centuries, this isolated island carnivore has managed to dodge anthropogenic extinction, unlike its less populous relative, the thylacine. And being twice lucky—

ironically now thanks to dedicated human intervention—the devil has seemingly also evaded the DFTD death sentence.

A 1927 newspaper in Oklahoma, USA, wrote that 'The "Tasmanian Devil" is one animal among the queer creatures of the Antipodes that seems to have a sufficiently characteristic appearance not to be given an animal name borrowed from foreign lands—at least earthly lands . . . He gives a faint suggestion of jackal wolf and wild cat rolled into one.'[1] Those observations neatly summarise the devil's twin people problem: a scary non-animal title and baffling identity.

Such was the animal's burden, albeit that some much earlier sympathy for the devil name had surprisingly been expressed, in England in the 1860s:

> We really think the animal boasting of the title of Tasmanian Devil deserves a name slightly modified in sound, as his appearance is anything but ferocious to the degree anticipated.[2]

Likewise:

> Intending visitors should not be scared by the repulsive appellation of this curious creature, for it is by no means a devil calculated to inspire dismay or dread.[3]

And:

> The 'Tasmanian Devil', to the disappointment of hundreds, was on inspection only an animal similar and about twice the size of an English rat.[4]

There is no indisputable original source for naming the animal as a devil. This 1899 description in an Irish newspaper nevertheless has a ring of truth: 'It is believed that the name of the devil was bestowed on the animals by the convicts, who had learned to look upon them with almost superstitious fear, partly in consequence of their appearance, but still more owing to their untiring perseverance in following up an enemy to the last with what looked like undying hatred.'[5] Could it be that this animal, being small, nocturnal and therefore 'invisible', yet making piercing unearthly shrieks in the night, was co-opted by early Van Diemen's Land gaolers as a fortuitous fright story intended to help them prevent their convicts absconding?

Intertwining such a name with convictism made sense to both the authorities and free settlers of the 'young' remote colony of Van Diemen's Land, that one was as bad as the other—untameable, irredeemable. This 1867 UK observation may seem amusing but underscores the point: 'The Tasmanian devil is sometimes eaten as a dainty by the hungry inhabitants of Van Diemen's Land, to which it was confined like some biped rascals who were transported thither.'[6]

Although historically well understood as (mostly) honest descriptions of their time, wild exaggerations excoriated the animal further. Here are some choice newspaper morsels:

UK, 1866:

An animal soon learns to recognise its keeper, and to welcome the hand that supplies it with food, but the Tasmanian devil seems to be diabolically devoid of gratitude, and attacks indiscriminately every one that approaches it.[7]

UK, 1868:

Like a dwarfed hyena . . .[8]

UK, 1870:

A certain weird and uncanny look . . . More obtuse to kindness than the hyena, more treacherous than the wolf, relapsing from paroxysms of purposeless rage into short lulls of sullen sulk.[9]

Australia, 1871:

The Tasmanian devil, that curious animal that with its powerful jaws is able to completely crunch up a sheep.[10]

UK, 1880:

A fierce sort of prehistoric badger.[11]

Australia, 1882:

A most repulsive-looking creature . . . It ought to be classed with the 'gorgons', 'hydras' and 'chimeras dire' of classic fable.[12]

UK, 1883:

Perhaps the most cruel, ferocious, and restless of created things.[13]

US, 1885:

The most utterly fearless thing on earth.[14]

US, 1909:

The devil looks like a wild hog that has been telescoped, losing his neck in the process.[15]

Australia, 1913:

In Tasmania the hand of every beachcomber and settler is against this snarling Ishmael of the scrub.[16]

There were attempts to rationalise devil behaviour, one being that, as a prominent island predator not prey species, 'his brain never seems to have acquired any proper organs for feeling fear' (1881).[17] This was in keeping with the broader Acclimatisation-era notion that marsupials were a distinctly lower order of mammals than placentals. While such an unlikely hypothesis might have seemed plausible—in the way that Antarctic and sub-Antarctic animals show little fear of humans—to accuse the species of 'unmitigated wickedness' surely undid any attempt to rationalise its behaviour. The direct satanic tag, once applied, could not easily be undone, even into the early 20th century:

> It seeketh after darkness and escheweth the light. By day it hides, meditating evil in the caves and waste places of the earth; at night it comes out, and walketh up and down the earth to and fro, seeking for prey . . . There appears to be only one way of taking advantage of these animals. Any bright light blinds them.[18]

Elsewhere in this book communal devil feeding is described. The animals tug at tough, sinewy, boned carcasses, breaking them up. Feeding devils look aggressive, but they are not. Yet those behaviours allowed for ample exaggeration: 'The hand of every farmer is against this ferocious animal, which kills more than it can devour. It is recorded of one which had escaped from confinement that in two nights it killed 54 fowls, six geese, an albatross, and a cat.'[19]

Demonising the devil needs to take cruelty into account. The treatment of animals in the wild and captivity was frequently horrendous, the latter being in zoos, travelling menageries or on board transportation vessels. An 1867 description of a caged devil stated that 'It frequently springs at the bars of the cage in which it is kept, and endeavours by dint of teeth and claws to render them asunder, at the same time giving vent to its rage in short hoarse screams.'[20] The true story of that particular animal, however, emerges by swapping 'rage' for 'abject terror'. As is this case of torture:

> The *Auckland Herald* reports that more 'wild beasts' and rare birds have arrived from Tasmania in the Bella Mary. We observe that she has on board a couple of opossums . . . Another attempt was made to introduce the Tasmanian devil [to New Zealand], but the brute pined to death in the apple case in which he was placed.[21]

In William Shakespeare's play *Hamlet* (1599–1601), the title character says, 'I must be cruel, only to be kind,' which can find many echoes in the Tasmanian devil. This 1925 UK zoo description is apt:

> Australia gives us an amazing caricature—this time caricature of a temper. It is owned by the Tasmanian Devil—and he is always losing it. To make the Tasmanian Devil happy you have only to give him something to kill. Kind-hearted keepers always do this when they clean out his cage. The victim given to the devil is poor 'Sammy.' Sammy is a long iron rod with its end bent into the shape of an 'L'. The door of the den is opened, Sammy goes to its doom between two locked jaws and perishes slowly while the keeper changes the devil bedding. Then the man has the painful task of recovering Sammy's remains. Were he to pull directly the rod the teeth and jaws would break before they released it, but a little sideways twitch does the trick and the 'L' slides clear out of the corner of the mouth from behind the teeth.[22]

Captive animals became big business, especially if they fell into the exotic category, which the Tasmanian devil certainly did. Displaying animals also meant describing them, which proved tricky:

> Now Exhibiting at Tweed's Public Gardens, Dirk Hill, Horton, Bradford, and can be seen Alive, The Wondrous Iconernosacus or Tasmanian Devil; head like the Hippopotamus, body like a Bear, claws similar to the Tiger, and ears similar to the Horse. The most eminent naturalists of the day have failed to trace the origin of this peculiar animal.[23]

That in itself proved part of the attraction—what could this creature be? As a newsletter associated with London's Royal Zoological Society noted in 1870:

> Within the past few days the council has secured a third Tasmanian devil . . . The number of visitors to see them since their arrival bears ample testimony to the wisdom displayed by the council in their anxiety to cater for the instruction and amusement of the friends of the Society.[24]

Twenty years later, London's Zoological Gardens had acquired another three devils and formed a distinctly more rational assessment of them:

> The popular name applied to these animals may seem to the visitor to convey a rather exaggerated idea of their ferocity. Whatever their disposition towards other creatures may be, they are very well behaved towards each other. All three may be observed curled up in friendly proximity upon the straw in their cage, only keeping a wary eye upon the casual passer-by. It has been stated that it is impossible by any amount of kindness to eradicate the sullen ferocity of these animals.

> The keepers at the Zoological Gardens have been so successful in establishing themselves on a friendly footing with such unpromising creatures as sea lions and so forth, that they may succeed even with the dasyures.[25]

Back in Tasmania, the animal's demonic status also shifted as its entertainment value was realised. In 1876, a Hobart advertisement boldly declared:

> THE REAL TASMANIAN DEVIL. This creature is peculiar to this Island alone, and is described by naturalists as THE WONDER OF THE WORLD. A PAIR will be exhibited at ELWICK during the Races.

In the 1880s, Moncure D. Conway, a US slavery abolitionist and radical thinker, visited Tasmania. He refreshingly called out the nonsense of previous generations:

> Wonderful is the little 'hand fish', which climbs up on the beach sands, props itself on its finny hands, and looks at one as pertly as a sparrow. The 'Tasmanian devil' is a good deal of humbug, too. At Auckland I heard him described as fierce, untameable, dangerous . . . at Melbourne he sank to 'an ugly little beast'; in Tasmania it is discovered that the poor little nocturnal creature is rare and timid, and will run for his life at the slightest noise![26]

There was belated realisation, too, of the benefits of the animal in the wild:

> Happily the worst enemies of the old-time squatter have been thinned down or exterminated—the dingo, the marsupial wolf, the Tasmanian devil—but probably the Colonists would gladly welcome them back could they get rid of the imported rabbits and of the sparrows.[27]

The resurrection, so to speak, of the Tasmanian devil continued through the early decades of the twentieth century. In 1909, devils were sent to the USA for the first time:

> A letter received from Honolulu by a friend of Mr. H. D. Baker, American Consul [in Tasmania], states that all the Tasmanian animals which he is taking to the United States have crossed the tropics in good health, and there seems every prospect of their reaching their destination in safety . . . The four Tasmanian devils . . . seemed to lose much of their savage demeanour towards outsiders as the voyage progressed. At first, if a stick was poked into their cage, they would open their mouths and snarl, and then snap savagely, but after a while it was impossible to get a growl out of them, and they would playfully gnaw at anything stuck into their cages. It seemed as if the general opinion that the Tasmanian devil is impossible of taming might be incorrect.[28]

This delightful story appeared in 1934 in Brisbane's *Telegraph*:

> Attendants at the [Melbourne] Zoo are smiling over an incident which occurred recently when it was discovered that one of the Tasmanian devils was missing from its enclosure. Two of these animals were removed recently from the small marsupial house to an enclosure next to one which is the home of several little white rabbits. A short time ago one of the Tasmanian devils was missed. A search was begun at once, because it was feared that it might kill some of the other small animals in the Zoo. Several men with dogs hunted in vain for the missing animal, but it was a keeper, who was going [on] his rounds, that found the missing Tasmanian devil. He was lying comfortably in the white rabbits' enclosure with a white rabbit snuggled up on either side of

him. Although the rabbit is the natural prey of the Tasmanian devil in his wild state, this 'devil' had not attempted to touch any of the defenceless little things in whose company he suddenly found himself when he wriggled through from his enclosure into their yard. Perhaps he thought that the snowy whiteness of these particular rabbits gave them a special dispensation from the depredations of devils. 'I have heard of the lion lying down with the lamb,' the director (Mr. Wilkie) said, 'but this is the first time I have heard of the devil lying down with a white rabbit.'[29]

7

EXPLOITING THE DEVIL

> These days, I live on a small private nature reserve in the Tasmanian highlands, where a whole family of wicked-looking though loveable black beasts regularly invite themselves to feast at my tent, sometimes around midday with the sun shining through their red ears, often in the dead of night, dressed as they are for darkness and cocktails . . . Some say Tassie devils are innately convicts, thieves and criminals, but I prefer to think they are nature's creatures of fortune, as boisterous and inquisitive as children, who enjoy each other's company, laugh at their own jokes, and share what they find. For what else would they have done with those five missing shoes, champagne bottle, and two billiard balls?
>
> John R. Wilson, Quoiba

A singular irony pertaining to the Tasmanian devil is the small human population of its home island: lutruwita (c. 40,000 years), also Van Diemen's Land (from 1803) and subsequently Tasmania (from 1856). Once devils began to be displayed beyond the island—in mainland Australia, Europe, the USA, Japan, Argentina and other countries—captive devils were seen by numbers of people vastly in excess of Tasmanians actually seeing them in the wild. Thus, long before apples, they became the island's most recognisable export:

> For generations before this [Second World] war, about a hundred million people at least in the United States were living in heathen darkness of the very existence of Tasmania. But George Fitzpatrick altered that almost in the twinkling of an eye. That great friend of Tasmania is, beyond doubt, a past master of publicity. The year before the war, he had the whole of the American press telling everybody everything about Tasmania. How did he do it? Well, he took ten or twelve Tasmanian Devils to the United States, and those strange, forbidding-looking little animals created a splendid uproar throughout America by burrowing their way to liberty out of the zoos there. They were nobly helped to make Tasmania famous by a press always feverishly in search of sensations and 'stories'. This sounds funny and of course it is. But the idea was a stroke of genius, and the results were highly important. In less than a month there was scarcely a man, woman or child in the United States who did not know all about Tasmania.[1]

Exploiting the demonised devil for commercial gain is nowhere more evident than in this ludicrous account, as reported in Launceston's *Examiner* newspaper in 1924 under the triple headline: 'A New Rejuvenator. Gland of Tasmanian Devil. An American's Theory.' The full article reads:

> The following is an extract from the *San Francisco Chronicle*. Bringing an assortment of what is believed to be the most powerful restoration glands in the world, T. Lawrence Braun, globe trotter and scientist, has reached San Francisco from

> Tasmania. He is *en route* east for experiments with the serum of the Tasmanian devil, the toughest animal known to science. Braun declares that the Tasmanian devil, which is a small, bear-like animal which lives in Tasmania, is almost impossible to kill. The animal's tenacious grip on life impelled him to experiment with its glands. He claims to have discovered some astonishing facts. 'The value of these glands first came to my attention,' he said, 'when I accompanied a shooting party on a trip to clean out these devils, which prey upon sheep. We cornered one in a narrow canyon. Five shots from a heavy carbine, while each one turned the animal completely over, did not seem to affect its desire to attack us. Finally a native literally beat its head off with a gun butt.
>
> 'The chief hunter said he would show me something. He cut the animal open, and, to my surprise, I found the heart was still beating. It continued to beat for nearly half an hour after the animal was, to all intents and purposes, dead. It struck me that such astonishing tenacity of its organs to continue their natural functions would result in gland secretions of unusual virility. I obtained the glands from a number of the animals, and later, with certain eminent medical men conducted a series of experiments in Paris.
>
> 'The results were astounding. My present trip to New York is for the purpose of demonstrating this new serum to a group of scientists, pursuant to the cultivation of the Tasmanian devils as the super-gland producers of modern times.' Braun declares that he values his present serum collection, contained in half a dozen small bottles, in excess of 500,000 dollars.

Unlike Mr Braun's marsupial snake-oil salesman pitch, the apple industry comparison is not fanciful. This is because the remote southern island became so identified with that fruit as to take on the moniker 'the Apple Isle'. Tasmania's burgeoning international identity became rather confusingly mired between godliness—the apple being the Tree of Knowledge fruit in God's Garden of Eden—and His satanic arch-nemesis the Devil.[2]

A devil sampling a product from God's Garden in an advertisement from The Daily Mail, *28 April, 1937.*

Nonetheless, both devil and apple became profitable products. Prior to cool storage shipping, Tasmania's apples didn't travel well overseas. It wasn't until the mid-1880s that the industry accelerated, becoming the island's second-most important export after sheep.[3] The devil, on the other hand, had been exported since at least the early 1860s (even if pining to death in an apple case from time to time).

So it was that, in the mid-1860s, a Cheshire, UK, newspaper noted that 'The Tasmanian Devil still proves both an object of attraction to the visitors to the grounds, and a source of revenue.'[4] A few years later, in neighbouring Lancashire, a Liverpool newspaper advertisement undoubtedly referred to the same devil:

> T. M. Fernyhough respectfully informs his Friends and the Public that he will open his tent during the races opposite the Grand Stand, with a Choice Selection of Wines, Spirits, &C. He will Exhibit the Live Tasmanian devil in his tent as above.[5]

The writer of this 1861 Sydney newspaper advertisement took a less polite approach:

> THIS EVENING.—The DEVIL! The DEVIL!! The DEVIL!!! The TASMANIAN DEVIL!!! To be seen at the London Saloon. Be in time to see the Devil. Admission, 6d.[6]

In 1880, Tasmania participated in the Melbourne International Exhibition, the first official World's Fair held in the Southern Hemisphere. The Royal Exhibition Building, built for the exhibition (and a World Heritage–listed building since 2004), hosted some 1.3 million visitors who experienced and discovered tens of thousands of products and new creations from all over the world. The Expo played 'a critical role in the development of Melbourne.'[7] In its reporting on the Exhibition, the *Mercury* newspaper observed:

> The Tasmanian Devil is a capital introduction to the Tasmanian Court, the wiles of the former having often the effect of alluring spectators, not particularly interested

in the products of Tasmania, to examine the varying and multitudinous exhibits of the latter . . . The devil was frequently enquired after, and certainly it now appears to receive a somewhat remarkable share of attention.[8]

The plunge in wild devil populations in the first decade of the twentieth century was noticed; hence, the devil's reputation in 1906:

The Tasmanian devil is in splendid health and will form a very valuable acquisition . . . the only specimen in Britain and the first to be seen in Scotland for over twenty years . . . In shape it somewhat resembles the badger, but is black in colour, with a white gorget on its chest.[9]

But what of an actual monetary value? The Bostock and Wombwell travelling menagerie, a year later, received this praise:

They bring to Cheltenham some of the greatest novelties on earth, headed by one of the strangest animals alive, and the only one of its kind in captivity—the Tasmanian Devil. For this animal Mr. E. H. Bostock is said to have refused the sum of £600 offered by the Zoo authorities, London.[10]

That's a good measure of real value, a far cry from some 30 years earlier, when a devil from the same menagerie went under the hammer:

Wombwell's Menagerie was submitted for sale at Edinburgh on the 8th March, in the Waverly Market. A very beautiful young kangaroo was sold to Mr Rice for £12, and an emu—a very suitable bird for a gentleman's park, and a nice show thing for the ladies in the morning after breakfast—was

> secured by Mrs Day for her collection at £7. An animal with the title of 'Tasmanian devil,' which the auctioneer assured his hearers was stronger in the jaw than an hyaena, although he would not recommend him for a house pet, was also submitted to public competition. Bids were slow, and even the prospect of purchasing the 'devil' for £3 did not render buyers enthusiastic, and Mrs Day bought him at an advance of 5s. on the sum.

The top price was an elephant, selling for £680.[11]

The devil traded as a regular Tasmanian commodity well into the twentieth century, until the then Animals and Birds Protection Board placed a ban on its capture. Here is just one example, from 1936:

> Shipping Information—Burnie—Exports, July 29.
> Wollongbar, s.s., for Melbourne—79 sacks potatoes, 236 sacks swedes, 50 sacks peas, 3 cases soap, 16 empty return casks, 1,650 super feet timber, case wireless goods, 4 bundles skins, 13 bags hides, 14 bags turnery, 3 bags rabbit skins, 33 boxes butter, 14 bags bacon, 6 hampers samples, 13 packages films, one Tasmanian Devil.[12]

Overseas export for commercial purposes—particularly zoo displays for fee-paying visitors—did not let up. Thus, in April 1938:

> The Australian 'Noah's Ark' has left Brisbane for America. It is the steamer Port Halifax which is carrying 789 Australian birds and animals in 73 large wooden crates . . . 200 live pigeons were shipped for the daily menu of the two Tasmanian devils.[13]

A few months prior to this US shipment, there had been devilish lamentation of sorts in the UK:

> Satan, the London Zoo's Tasmanian Devil, has died . . . He was the last survivor of a batch of four which came to the Zoo six years ago . . . Although Satan will be unmourned, he will be greatly missed . . . as these creatures are becoming very rare.[14]

Alas for that zoo. Within a year, Hobart's *Mercury* newspaper reported:

> 'Dammit', the London Zoo's last Tasmanian devil, is dead. His disappearance is much regretted, as it is realised that he will be most difficult, if not impossible, to replace. 'Dammit' was the survivor of four 'devils', and was originally provided with three 'wives', who have died one by one. They all failed to reproduce their species in captivity. Age did not improve 'Dammit's' temper, and he had his neighbours in the surrounding cages in fear of his tantrums, even until almost the end.[15]

Exploiting the devil's 'evil' nature became a lure of sorts, as evidenced in this 1906 menagerie advertisement:

> The greatest and most profound novelty ever seen in England, The Tasmanian Devil, alive. This Creature is never seen in any other part of the world except Tasmania, and is a terror to everyone. To be seen at Wombwell's, free of any extra charge, and no danger.[16]

A distinct commercial risk (unlike imported Codling moth disease, which nearly destroyed Tasmania's apple industry in the 1880s) was the devil's exceptional ability to escape from confinement. Here is 100 years' worth of what might have been labelled in those times:

AMAZING TRUE DEVIL ESCAPE STORIES!

1833:

When Captain Petrie lately sailed from [here] in the Drummore to the Mauritius, he took with him as a present to one of the authorities at Port Louis, one of those savage creatures peculiar to this Island, commonly called a devil, which had been caught and as far as was practicable tamed, by Mr. Davidson at the Government garden . . . But the devil was not long in his new berth when he contrived to make his escape, and for some weeks dreadful havoc was played among the poultry around, until with difficulty he was shot.[17]

1867:

[Adelaide Botanic Garden] obtained fine specimens, male and female, of the Tasmanian devil *(Diaboltus ursinus);* but unfortunately both have escaped—one from the Port and the other from the Garden. As these animals are becoming very rare, it is hoped they may be recovered . . .[18]

1874:

In 1867 two Tasmanian devils (*Sarcophilus Ursinus*) escaped from the [Adelaide Botanic] garden. One was shortly afterwards seen in one of the gardens in Glen Osmond, but after that nothing more was heard till the other day, Mr Schocroft, of Mount Lofty, informed me that he had caught a very curious animal in a trap, never before seen by him or the neighbours in that locality, which he would present to the garden. I was not a little surprised to find that it was a Tasmanian devil, and no doubt one of the deserters from the garden which had lived six years in the neighbourhood of Mount Lofty.[19]

1868:

The Tasmanian Devil had arrived at the exhibition at Bodmin on the previous Tuesday from Mr. Tamracke's, Ratcliff Highway, London, and was put into a compartment adjoining the crocodile, when during the night of Tuesday he ate through the partition, into the compartment with the crocodile, and a terrific encounter must have ensued, as he had torn the throat of the crocodile away; also broken up the bottom of his den and made his escape. The proprietor offered a reward for the re-capture of the animal.[20]

1868:

Escape of the Tasmanian Devil. This wonderful animal, the chief attraction of Wombwell's menagerie, effected its escape in the latter part of last week, and up to Saturday night had not been discovered. The menagerie was being exhibited at Bideford, North Devon. The animals had just been fed in the presence of the spectators, and the establishment was about to be closed, when loud shouts rang through the place that the 'devil' had escaped. The man who fed it omitted to fasten the door of the cage, and while he was getting a shutter the animal took advantage of his absence by leaping among the crowd and into the town. The excitement was immense. The animal was chased along the quay side and it boldly plunged into the water. Boats were immediately launched, and a diligent search made for the devil, but it was quite dark, and he could not be found. This is the third escape of the animal during his captivity—the first occurring at St. Day, in Cornwall, and the second at St. Just, in the same county. It is believed that the animal was drowned. It was of great value to the proprietors of the menagerie. Accompanied by some of the inhabitants of the town, they continued the search till Saturday night.[21]

1868:
Wombwell's 'Tasmanian devil' which escaped about a fortnight ago at Bideford, in Devonshire, has been recaptured in North Yorkshire. It was taken on a farm at Fremington; not, however, before it had made sad havoc with the poultry.[22]

1872:
The she-imp which so mysteriously disappeared from the [Hobart] Poultry Show was found safely lodged in the recess under the grate in the Town Hall, her majesty having alike disdained the attempt to escape by the chimney or conceal herself in the organ. Her cub, however, is still missing . . . Great consternation was created when the Town Hall doors were reopened, for the devil and her little imp had escaped through the iron-barred door of their apartment. Several pieces of wire, bent and bitten, as well as chips of pine, were scattered about the floor of the hall, but the whereabouts of the fugitives could not be discovered . . . The sleek and sooty little *Sarcophilus*, which was missed from its mother's back in the Town Hall, was discovered next morning by the hall keeper. She was evidently tired of the growling of the organ, and appeared much weakened by three days' fasting. She soon rallied, however, under the tender care of the landlady of the Hobart Town Hotel, who provided warm milk and other nourishment for the famished runaway.[23]

1873:
LOST, Tasmanian Devil, with collar and chain on. Finder rewarded. 4 Royal-lane, Bourke Street E [Melbourne].[24]

1875:
By an unforeseen accident (says a New Zealand paper) the Tasmanian 'devil' on board the Bella Mary got loose and

escaped to the hold of the vessel. The crew were mustered on Sunday to capture the 'devil', but after many futile attempts the brute managed to escape by bounding over the head of the chief officer, and being hotly pursued, took flight by jumping overboard. It is uncertain in what direction the animal took its course. He appeared to be an expert swimmer, and struck out boldly for the North Shore, but is improbable that he would have continued his course so for a long distance. Darkness prevented those on board the Bella Mary from observing the 'devil's' movements; but it is to be hoped that he has met with a watery grave.[25]

1881:

The proprietors have been rather unlucky since their arrival in Launceston in regard to their animals. A short time ago they purchased five native devils, and secured them all together in a strong cage with iron bars. However it appears that in the case of the Tasmanian devils 'Stone walls do not prisons make, nor iron bars a cage', for to the surprise of the proprietors upon entering the tent yesterday morning they discovered that three of the number had effected their escape by squeezing themselves through an aperture between one of the end bars and the woodwork of the cage. Search was made for the missing ones, but up to last evening these three devils were still loose, and are doubtless at present enjoying that freedom allowable at this festive season of the year.[26]

1882:

The Zoological Collection. A letter to Alderman Farrelly from Dr. R. S. Smythe, of Latrobe, was read, stating he had forwarded a Tasmanian devil for the People's Park, and had a much larger one, but it had escaped.[27]

1882:

It is said a travelling menagerie which visited the district a few weeks ago lost, at Ulladulla, two Tasmanian devils. These have never been seen, but as they have not been found dead it is supposed they are still in the neighbourhood, and able to secure prey to keep them alive.[28]

1884:

This fine collection of [Bostock's Menagerie] wild animals paid a visit to the village on Wednesday last. Shows of any kind are of rather rare occurrence here . . . By some unforeseen circumstance one of the animals—the 'Tasmanian devil' as it is called—got out of its cage and escaped in the course of the evening. The fact of such a ferocious animal (which is said to be of the tiger cat species) being at large has . . . scared a whole county.[29]

1885:

A Tasmanian devil escaped from the [Launceston] Museum a fortnight ago, and has since been at large. This morning as Superintendent Propsting, of the Territorial Police, entered his office he was surprised to see the devil sitting in his chair. The creature was quickly captured and returned to the Museum.[30]

1892:

LOST, from Victoria Brewery, E. Melb., Tasmanian Devil, getting bald-headed, answers to name of Tom. Finder rewarded returning same . . .

. . . We have heard of a man who made a pet of a Tasmanian devil—one of the most obscene-looking animals in creation.[31]

Bipedal young devil (below) and wolverine (top)—the similarities in these unrelated mammals are striking. (Wolverine: Daniel J. Cox, Natural Exposures Inc. Tasmanian devil: *The Mercury*)

A young adult devil. This backlit photograph was taken in the wild. (Christo Baars)

A pair of adults establish the feeding hierarchy at a food site. (Christo Baars)

This devil was photographed by a geologist working in the remote and seldom-visited Wilson River area in Western Tasmania. The geologist probably surprised the devil, who in response climbed the tree. Tasmanian devils are efficient climbers, swimmers and diggers and are born to run; they are tenacious, versatile hunting scavengers. (Authors' collection)

Wildlife biologist Phil Wise with the first devil born on Maria Island. (Sarah Peck)

Ollie and Donny. Donny and Clyde were orphans raised at the home of David Pemberton, his partner Rosemary Gales and their children Sam, Elsa and Ollie. (David Pemberton)

Taz, the indomitable Warner Bros. character. Just five Taz cartoons were made between 1954 and 1964. He was resurrected in 1990 and became a billion-dollar income generator for Warner Bros. (Courtesy Warner Bros. Taz, Tasmanian Devil and all related characters and elements are trademarks of and © Warner Bros. Entertainment Inc.)

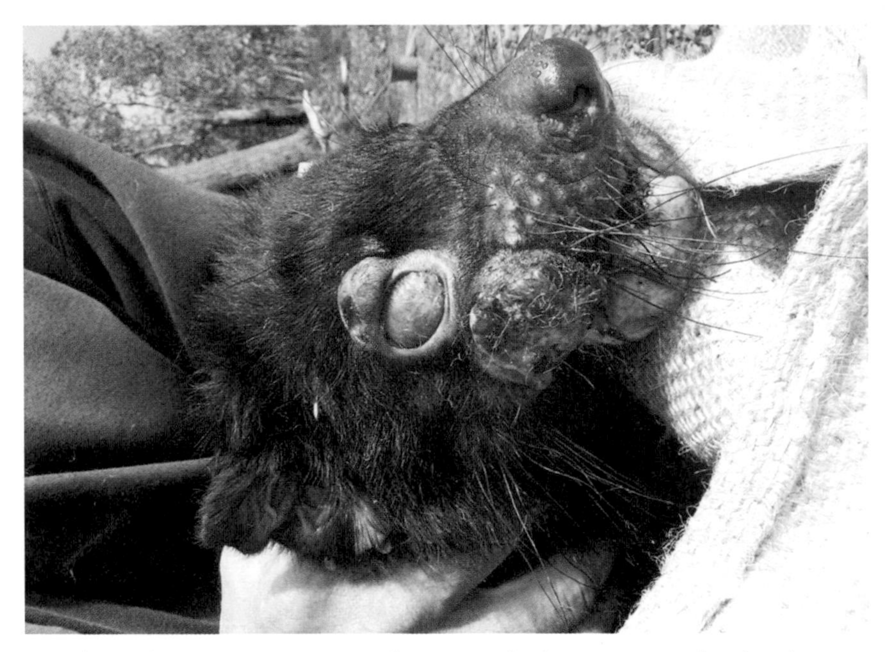

Devil Facial Tumour Disease is a virulent cancer of unknown origin and with no known cure. First detected in 1996, it has spread across most of the island, greatly reducing the devil population in some areas and threatening the survival of the species. (Nick Mooney)

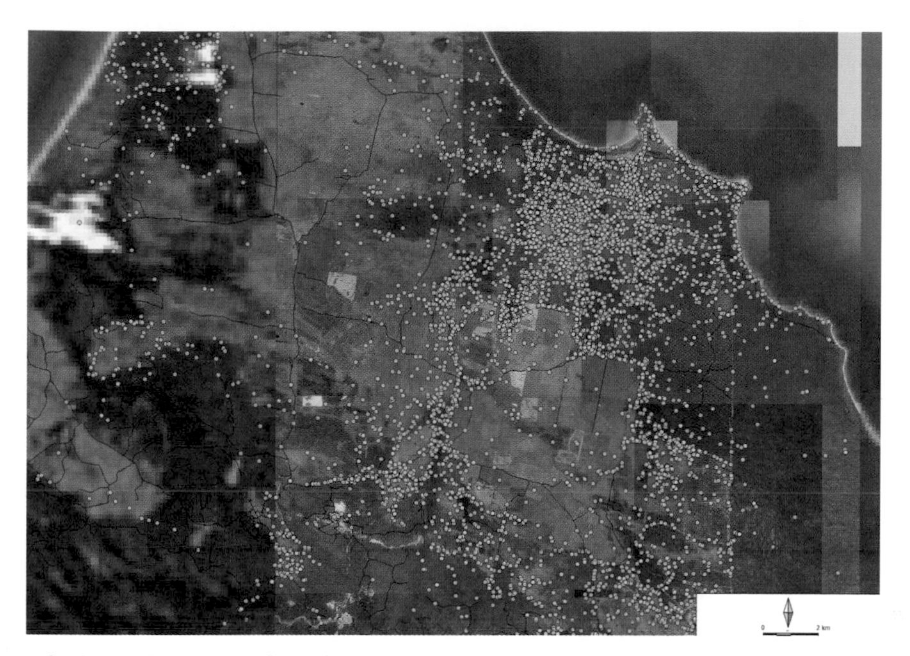

The dots on this satellite photo of wukalina/Mount William, in northeast Tasmania, represent the location of Maria Island devils released into the wild. Almost all of the devils settled close to their release sites. One exception was a male who in two months travelled over 700 kilometres within the area shown. (Devils typically avoid open ground for its lack of protective cover.) When examined by researchers, that devil had put on weight despite his travels. (Courtesy Department of Natural Resources and Environment Tasmania)

Opposite: *Aunty Patsy Cameron, a highly regarded Elder, author, researcher and cultural historian, made this basket to represent a devil den that was entwined in the roots of a peppermint gum in northeast Tasmania. The roots bind the den in the ancient soft sands of the dunes and provide refuge for young devils left behind when their mother leaves to forage. Aunty Patsy explains that this tree is the totem tree of Pairrebeenne/Trawlwoolway leader Mannalargenna (d. 1835) and the link to long-term devil dens. The basket was made from grasses found in the vicinity of the den.*
(Collection Tasmanian Museum and Art Gallery)

Eric Guiler initiated modern research into the Tasmanian devil; he is seen here enticing a devil to 'sing'. He also devoted much of his career to searching for the thylacine. (Courtesy *The Mercury*)

A devil den and latrine site at the base of a sandstone cliff at Fentonbery. High quality dens such as this probably remain in use for hundreds of years. (Courtesy Billie Lazenby)

1906:

The *Omrah* landed at Plymouth last month a Tasmanian wolf and a Tasmanian devil. When the vessel was a little way out and in Australian waters the devil, which is a carnivorous animal about 18 inches long, escaped from its cage, and defied all the efforts of the crew and passengers to capture it, being finally lost to view on the upper deck. A diligent search was made, but the animal hid away for three weeks. It was found at last, hidden under some gear on the upper deck. Evidence was not wanting that it had meanwhile been wandering about the ship by the fact that it was much fatter than when it escaped.[32]

1923:

After several attempts extending over a week, the two Tasmanian devils that escaped from their enclosure in the Beaumaris Zoo on the Domain [on its grand opening day] were recaptured on Friday night. The animals, possibly frightened by their strange quarters, tore their way through the strong wire netting of their cage, but did not escape from the grounds, and took refuge in a culvert at the southern end of the gardens. Baits were laid for them on several occasions, but although they were taken each night the animals could not be caught. Mr. Reid, the curator, constructed a box trap and set it at the mouth of the culvert on Friday evening, a very tasty bait of lamb's heart being set in order to induce the escapees to enter. This they did, and on Saturday morning they were back in their reinforced enclosure.[33]

1946:

In Melbourne police cars were called by radio to hunt for two Tasmanian devils which escaped from the zoo at Royal Park. The police were warned that the animals were harmless left alone. However, they might attack anyone who tried to pick them up.[34]

1946:

Temporarily housed at City Park Launceston are three Tasmanian devils which are destined for Taronga Park, Sydney. Originally there were six. Five escaped by burrowing under the concrete floor of their cage, and two were recaptured.[35]

1954:

The Tasmanian devils . . . are far from affectionate. Four, which arrived in the consignment of Australian animals, started to brawl as soon as they were put in their new run. Then one of the females burrowed under the wire and escaped. She was found half a mile away, having crossed a major road. This, however, is nothing compared with the record another Tasmanian devil put up a year or so ago. After having been missing for a week, he was found intact in Baker St., which is one of the busiest thoroughfares in London.[36]

1970:

Beware of the Devil. She may roam your chicken yard tonight. This she-devil may not lead you into temptation, but could sorely try your temper if she is tempted by your chickens. Fears are held for the safety of chickens in the Lincoln Springs [South Australia] area. A young Tasmanian Devil has escaped by slashing through the wire of her enclosure, and with its taste for chooks is expected to slash her way into a chicken run just as easily. The young Tasmanian Devil is owned by Mr. and Mrs. R. V. Provis. They had bought the Tasmanian Devil, a female, in Port Pirie for about $50, and were planning to mate her in the near future. However, grub came before love and the little devil broke the engagement by deserting the love nest.[37]

8

LURE OF THE DEVIL

The old 12th Battalion considered they had the most unique mascot of all A.I.F. units. It was a Tasmanian devil, and, if there is one animal in this old world appropriately named, it's the Tassy devil. He always appeared to have a liver, and, unlike most animals, he couldn't be bribed into a reasonable frame of mind. One after the other tried his hand at taming it, only to give it up in disgust. So eventually it was left to the tender mercies of a babbling brook [cook], one Bluey Thompson, of 'D' Company. Although the little devil was so unsociable, one and all of the 12th were proud of it. And all visitors [in Cairo] were paraded to look at this native of the tight little isle.

AJAX, WORLD WAR ONE 12/52 BATTALION VETERAN,
AS QUOTED IN THE *WESTERN MAIL* (PERTH) 5 NOVEMBER 1936

A later chapter describes the association of the Tasmanian devil with Warner Bros. and their billion-dollar-value cartoon character, one unedifying spin-off of which became an eight-year fight between the entertainment giant and a small rural Tasmanian manufacturer of a humble fishing lure called 'Tasmanian Devil'—a term that Warner Bros. had copyrighted.

The 'lure' of this chapter is that the real animal has irresistibly hooked the attention of millions since it first came to European notice as an unspeakable night demon, lying in wait for absconding convicts (though tasting like veal).

Intriguingly, although the television cartoon character first appeared in 1954 (and the copyright stoush came much later in the 1990s), a Warner Bros. connection had been established considerably earlier, in September 1939—the month of the outbreak of World War II. It happened this way, as reported by the *Los Angeles Times*:

> Look out! A couple of Tasmanian devils are roaming Griffith Park. Superintendent Gilbert L. Scott so reported to the Board of Park Commissioners at its meeting yesterday. 'They just gnawed through the wire cage, I guess we underestimated their capabilities. It's a devil of a situation,' he added. The black-and-white teddy-bear-like animals were presented to the Zoo last Saturday by Priscilla, Lola and Rosemary Lane, actresses, to whom a mother devil and two offspring were sent from Tasmania as a goodwill gesture. The offspring escaped.[1]

This (mysterious) 'goodwill gesture' soon inspired a headline: 'Hollywood Goes to the Devil!' Readers were informed that:

> It is not meant to imply that the capital of make-believe is deteriorating, but to suggest that Hollywood got a 'devil' of a kick from the arrival of three Tasmanian Devils consigned from Australia to three lovely stars of the motion picture constellation—Priscilla, Rosemary and Lola Lane. The Warner Bros. 'sister act' got time off from making 'Daughters Courageous' to receive the devils, but soon found that the belligerent little fellows couldn't actually be kept around

the studios as pets. These representatives of the island State's natural fauna have a nasty habit of chewing things up in their powerful jaws and, as nobody wanted to see Priscilla, Rosemary or Lola lose a leg, the devils were handed over to the Los Angeles Zoo at a special ceremony attended by the stars, the Mayor of Los Angeles and Tasmanian businessman George Fitzpatrick. Originally, the ill-tempered shipment from Australian shores was supposed to consist of but one marsupial mammal for presentation to Rosemary Lane—but during the month the devil was on the high seas she acquired a family, thereby saving friction in the Lane camp. 'Needs must when the "devil" drives,' Priscilla wisecracked after the devils had been safely quartered in their cage at the Los Angeles Zoo. She and her sisters were wanted back at the Warner studios to complete the day's shooting on 'Daughters Courageous'.[2]

The Tasmanian devil played a role in both world wars:

About 100 members of the A.I.F. Originals' Association fought their campaigns again when they got together at the annual reunion at the Launceston Cricket Ground on Saturday night. The men, about 40 or 50 of whom were from the South, sailed on the Geelong and Katuna from Hobart for Gallipoli in 1914. On Saturday night, the men lived again the days of the Gallipoli and French campaigns. One of the best stories of the night was of 'Bluey' Thompson, the best cook the army ever had, according to his former customers. 'Bluey' lost his pipe while on active service, and it was claimed that it was not found until some months later at the bottom of one of the dixies used for making stew. He also is credited with taking a Tasmanian Devil away with him on the *Geelong* . . . which was kept at the end of a long chain in the open-air kitchen. The Tasmanian declared if they caught the Kaiser they intended to shut up [the devil] in a cage with him and let them decide the rest of the war by single combat.[3]

Bluey's devil was eventually given to Cairo Zoo. And in 1941:

The first fully trained R.A.A.F. men to leave Australia have arrived overseas. They are all officers, and include pilots, observers, and gunners. They were members of several contingents of R.A.A.F. and A.I.F. men and nurses which have reached their destination. Nurses on one transport cheered the R.A.A.F. men as they embarked, and sang 'For They Are Jolly Good Fellows'. At another wharf, a train carrying Tasmanians had scribbled on it: 'Tasmanian Devils for Export.'[4]

The lure of the devil has also long attracted journalists, shaping public fascination with the species into memorable prose. Some examples:

1875:
We have heard much talk lately of His Satanic Majesty . . . If he is the original of man, then unquestionably civilisation has done something for the race, for this brute, making his escape once near Taunton, visited the fowl-houses of the locality and gobbled up the roosters, feathers and all, before they could give their owners warning by a single 'cock-a-doodle-doo'.[5]

1883:
He holds forth at sunset in a kind of falsetto, which sounds like a lot of schoolboys squabbling at play, and, in fact, we mistook his noise for that until enlightened.[6]

1883:
Sir—having just received [in London] a letter from Hobart,

upon glancing at the corner of the envelope, my attention was attracted by a peculiar postage stamp. Being in the company of several gentlemen at the time, I handed it round for inspection . . . One gentleman, whose education in natural history had evidently been neglected, suggested that it was a picture of the Tasmanian devil, and the protuberance from the head represented the horn. It was in vain I asserted that we had an animal in Tasmania called platypus, and described its peculiar formation. They evidently considered I was taking advantage of their ignorance, and mournfully departed to spread the intelligence that we Tasmanians had enthroned the devil in the place lately occupied by our Most Gracious Majesty the Queen.

Yours, etc.[7]

1896:

Its coat looks like wool rubbed the wrong way, and the general appearance of the beast is of a kind that makes avoidance appear the wisest policy.[8]

1907:

Amongst the 'novelties' is the great 'Tasmanian devil', the only one of its kind held in bondage for the delectation of the sensation-loving public.[9]

1922:

Outside the small cat house at the London Zoo is a mysterious cage, apparently empty, yet generally well provisioned with meat. It is the house of Smut, the Tasmanian devil freshly arrived from Australia, and the only one of its kind in the gardens. Acting up to his name, Smut is a true spirit of darkness, never venturing forth from his box-retreat, in daylight, a habit which, unfortunately, debars visitors from studying this terror of Australian farmers.[10]

1927:
The Tasmanian devil is a half-developed primitive beast with more fangs than sense.[11]

1935:
Their idea of happiness is an unending fight.[12]

1945:
Dealers in antiques would be interested to hear of the Launceston City Council's pipe-organ in the Albert Hall. There are points about this organ which place it in a different category from most others. It has the ability to startle its audience and disconcert the organist. You never know whether it will respond to the player's caresses, or stay silent. It chooses most inconvenient occasions to remain mute . . . Its sound rumbles round the stage, where it is thoroughly muffled before it reaches listeners in the hall . . . Before being assembled in the Albert Hall at an important exhibition, this organ was in the Mechanics' Institute, where it was placed on arrival from England in 1860. Soon after it was installed it was the focal point of a singular happening. A Tasmanian devil escaped from a menagerie in City Park, and disappeared mysteriously. Peculiar sounds from the organ in the institute led to investigation—and there was the devil. Cynics declare the ghost of the devil has returned to haunt the organ in its old age.[13]

1951:
When he produced the Tasmanian devil, complete with dog chain and collar, in the bar at a New Norfolk Hotel [Tasmania] on Saturday, he was given plenty of elbow room.[14]

> 1958:
> What's a Tasmanian Devil? It's a foreign little dog that's furious all the time. What's he so furious about? Because he lives in Tasmania and can't go to Sabella's [California] for their special weekday lunches . . . Your favourite meat and seafood dishes (salads too) for only $1 to $1.50. Pretty infuriating to miss that, all right.[15]

On another culinary note, while hungry convicts were noted as eating Tasmanian devils for their veal-like taste, the animal has not been eaten by humans for a very long time now. But in the early 1950s, slowly emerging out of post-war austerity, Australians were able to consider savoury delights such as 'Sardine Specials', 'Crabmeat Canapes'—and 'Tasmanian Devils':

> Remove the stones from large soft prunes. Stuff with tomato chutney. Wrap each prune in a piece of streaky bacon which has been lightly spread with mixed mustard. Fix with a toothpick, or string on metal skewers. At serving time, bake or grill until the bacon is crisp. Turn them once for even crisping. If they have been cooked on skewers, slip them off and insert a toothpick in each. Serve immediately.[16]

This is a far more elegant porcine link than the 1899 description of a Tasmanian devil's head being 'shaped like a pig's with the turned-up end of the snout cut off'.[17]

9

IN THE MATTER OF THE SOCIETY AND THE BOARD

A ranger's wife learns to cope with all sorts of emergencies, but the most bizarre was the day I bathed a devil in the laundry tub. He had fallen into the sewerage ponds at Strathgordon. A Hydro worker had watched him swimming frantically around. He was eventually fished out with a pole, exhausted, panting, not looking (or smelling) very good . . . He stayed still, just shivering while I stood him in the warm water and lathered him up, working fast while my luck held. Lifted him out, towelled him down briskly, while he continued to look stunned, then relocated him to the shed, into a wooden box filled with warm blankets, and left him in peace to recover from his ordeal. When the children came home from school we went to the shed, opened the door and the devil bolted to freedom

MAUREEN JOHNSTONE, RIVERSIDE

Eric Guiler's pioneering research into devil populations remains of central importance. As a reader in zoology at the University of Tasmania, and chair of the state's Animals and Birds Protection Board in the 1950s and 1960s, he was uniquely placed to study the island's ecology. He provides an interesting potted modern history of the devil:

> This creature did not figure in the debates of the early Boards because, at that time, it was rare to uncommon over all of the State. It was not until 1945 that devils appeared in the Minutes when the Ranger at Lake St Clair was reprimanded for being knowingly involved in the capture of two (or more?) for Poulson's Circus.
>
> By 1950 the numbers had built up to enable the Board to grant permits for their capture and by 1959 pressure was being exerted on the Board to place them on the Unprotected Schedule on account of the damage they did to possums caught by trappers and to weak sheep and lambs. No action was taken other than to grant permits to the complainants provided they could prove the alleged damage.
>
> The depredations became more widespread and by 1966 the Board was issuing poisoning permits to control the devils. However, it was clear that very little was known about devils and a research project was started in 1966 in co-operation with the Zoology Department of the University.
>
> The Board was fortunate in that it had the resources at the time to commence this work as some of the Members were very reluctant to issue poisoning permits for an indigenous species about which so very little was known.
>
> The project was important as it showed that the Board was prepared to switch its resources into non-commercial or non-sporting species and treat them on their scientific merits rather than the political desirability of being seen to be studying the so-called game species. This was the Board's first research programme into a non-game species.[1]

Every year between 1966 and 1975, at remote Granville Harbour on the west coast, Guiler led a devil research team. They set more than 5000 traps at nineteen locations, with such names as Harrison's Back Pockets, Dead Heifer, Duck Creek Track, and Pig Farm. A total of 282 devils was captured 946 times, 664 being recaptures. He also carried out extensive field research at tebrakunna/Cape Portland in north-east Tasmania. The published results initiated modern devil research.[2]

But Guiler didn't have things all his own way. The politics of wildlife conservation and management is as robust in Australia as anywhere else, as he had earlier discovered through the seemingly innocuous and modest trade in zoo animal exports. For six years conservationists repeatedly attacked the Animals and

Eric Guiler (right) with Eric Reece, a former Premier of Tasmania, at the launch of Guiler's Thylacine: the Tragedy of the Tasmanian Tiger in 1985. Eric Guiler devoted much of his career to studying thylacines and Tasmanian devils. (Author's collection, Tasmanian Museum and Art Gallery)

Birds Protection Board for being a key player in the state's fauna export trade.

The saga began in 1957 when the Sydney-based Wild Life Preservation Society of Australia, through its journal *Australian Wild Life*, wrote to the Board expressing concern at the Board's decision to give two Tasmanian devils as a gift to a Swiss zoo. Guiler replied that the director of the Basel Zoological Gardens, Dr E. M. Lang, had specifically travelled to Hobart to supervise the safe travel of the animals, and that the devil population had increased to such an extent in some parts of the state that 'we have had to make special arrangements to reduce the number owing to their depredations'.[3]

Much of the June 1962 issue of Australian Wild Life: Journal of the Wild Life Preservation Society of Australia *was devoted to the Tasmanian devil and the apparent lack of regard for it by officials of the island state. The New South Wales-based Society's efforts on behalf of the animal may have inspired Tasmania's Animals and Birds Protection Board to commence research into devil populations. (Courtesy Wildlife Preservation Society of Australia Inc.)*

But the Society persisted, querying the legality of the deal and also questioning Lang's proficiency in wildlife management. The journal quoted a report in *The Mercury* in which Lang planned to return to Australia for more exhibits: 'They include the Lyre Bird and Koala Bear, which he is confident can be trained to live on a diet other than gum leaves, which are unprocurable in Europe'.[4]

The fur really began to fly, in the matter of the Society and the Board, with the publication of the June 1962 issue of *Australian Wild Life*. Using a photograph of a demure, attractive juvenile Tasmanian devil on its cover, the journal ran a lengthy article attacking Tasmania's slackness in not protecting its prime marsupial carnivore. The little animal, still a nonentity in the public consciousness, had now become unique and possibly endangered, yet the Board seemed careless of its welfare.

The long article is entitled 'That Devil Again', and begins with a report on an escaped devil from a New South Wales circus, refers scathingly to freezing devils in London Zoo, and concludes with a salute to a hunting magazine, *Australian Outdoors*, uncharacteristically deploring pastoral ignorance of Tasmania's unique little carnivore.

The article's centrepiece, however, is the reproduction of a lengthy letter written to Guiler by Thistle Y. Stead, the Society's Honorary Secretary, who was also editor of *Australian Wild Life.* Her letter included 21 questions:

1. Is the Board conducting a survey on the distribution and population densities of the Tasmanian Devil (*Sarcophilus harrisii*)? If so, when did the survey commence; what areas have been surveyed; and how many investigators are employed in the investigation?

2. Has the Board, or any member of it, published any matter concerning the distribution and population densities of the Tasmanian Devil? If so, where and when was such material published?
3. How many members of the Board are actually engaged in field research and have an intimate knowledge of the Tasmanian Devil in its wild state? What are the names of the members claiming such knowledge?
4. Has the Board made public any knowledge that it may possess on the distribution and population densities of the Tasmanian Devil through the Tasmanian Press or through questions asked in Parliament? Has it made public such knowledge in any other manner?
5. Has the Board made the public aware that though Tasmanian devils and other wholly protected fauna may be accidentally trapped, that it is ILLEGAL on the part of the trapper to transfer such animals as are trapped to places where they may be artificially confined under conditions opposite to their natural environment? If the Board has made this matter apparent to the public, by what means was it made apparent, and has the Board's opinion on this matter been published in the Tasmanian Press? If so, when?
6. In view of the obvious trend of the Tasmanian public against zoos, does the Board intend to issue further permits for the establishment of further zoos?
7. How many permits have been issued by the Board for zoos in Tasmania in the last 21 years? How many zoos are extant; and what are the precise localities in which they are extant; and what are the names of the holders of the permits?

8. In view of the national trend against trading and trafficking in wild life, does the Board intend to issue further permits to individuals likely to engage in such traffic?
9. How many permits have been issued by the Board for the export of Tasmanian Devils to all sources; and precisely how many Devils have been exported in the past 21 years?
10. In what year were the most Devils exported?
11. In instances where the Board issues permits to wild life traders or persons engaged in the exhibition or traffic of wild life—does the Board charge a fee for such permits? Has the Board fixed any limit to what may be paid or received in transactions dealing with the sale of Tasmanian Devils? If so, what is the monetary difference between the fee charged and the profit made by the individual trader? If so, how much is the fee for such a permit? If not—to the aforementioned—why are permits issued to individuals to possess wild life which is public property? Does the Board agree or not agree that a negative view is antipathetic to the purpose for which the Board was created?
12. Does the Board issue permits for the destruction of Tasmanian Devils which are public property, gratis—or is a fee charged? If a fee is charged for a permit, what is the monetary difference between the value of the fee, and the cost of inspection to the public to establish the rightness of the permit? If a fee is not charged, why is it not?
13. How many permits have been issued by the Board for the destruction of Tasmanian Devils since 1957?
14. Does the Board require the holders of permits for the destruction of Tasmanian Devils to return to the Board the numbers of animals destroyed? If not, why not?

15. In view of the scientific value of destroyed Tasmanian Devils, has the Board made provision to see that such specimens are not wasted? Is there a legal obligation on the part of permit holders to return destroyed animals to the Board?

    (The WILD LIFE, PRES. SOC. of AUST. is vitally concerned with this question in view of the thoughtless wastage of destroyed thylacines in the past, and the consequent rarity of both skeletal and anatomical material, which, for the most part is housed in foreign institutions. It is strongly urged that such a situation should not be permitted to occur in respect of *Sarcophilus*.)
16. Is the Board able to name Australian scientific institutions which have benefited by donations of specimens of *Sarcophilus*? If so, how many institutions have received such donations through the Board?
17. Is the Board able to name foreign institutions which have benefited by donations of specimens of *Sarcophilus*? If so, how many donations have been made in the past 21 years, and what are the names of the recipient institutions?
18. In connection with the press statement made recently by the Chairman of the Board concerning the alleged destruction by Tasmanian Devils at Bridport, would the Board state how many permits for the destruction of this marsupial have been issued to date, and how many animals have been destroyed?
19. Similarly, would the Board state how many permits it has issued for the destruction of Tasmanian Devils during the past three years at or near the following Tasmanian townships: . . . (a) Temma, (b) Marawah, (c) Trowutta, (d) Redpa, (e) Christmas Hills, (f) Mole Creek, (g) Bridport, (h) Hamilton?

20. Is the Board able to state what may be favourable variations in *Sarcophilus* tending towards its survival in satisfactory numbers? Similarly, is the Board able to state what may be unfavourable variations tending otherwise?
21. To state that one species of mammal is 'as common as any other marsupial' is too vague for an empirical and analytic appreciation of a species population disposition and position. Would the Board state more precisely the numerical relationship of the Tasmanian Devil to other marsupials?[5]

Guiler commenced his research less than three years later and it is likely this correspondence provided some impetus for it, given the apparent lack of official and scientific interest in devils at that time.

Australian Outdoors, a classic hunting and fishing magazine complete with advertisements for the latest firearms, boats and fishing tackle, ran an article in its November 1961 issue entitled 'Protection That Doesn't Protect'. The 3000-word article, by Jack Bauer, attacked Tasmania's farmers and graziers, accusing them of endangering the devil's future. Bauer wrote with impressive understanding of the animal, based on personal experience, and it is likely that this was the first non-scientific published account detailing the ecology and behaviour of the animal in the wild. (Bauer noted that: 'There is very little known about the animal. The only reliable information on this animal is given by [Ellis] Troughton in his classic *Furred Animals of Australia* [1941]. He gives 58 lines on this animal. Not much to go by.')[6]

Bauer's article led with praise for the protected status of the devil, while deploring the agricultural sector for branding it a ferocious killer of poultry and lambs. A farming group or groups had apparently been pushing the scare for some two years, and had now warned that if left unchecked the devil would become as great a menace as the dingo was on the mainland. Bauer insisted that those farmers were quite wrong, and that he had scientific proof to back him up. Proof from who? None other than Dr Eric Guiler of the Zoology Department, University of Tasmania.

Guiler had recently conducted a study of the stomach contents of eighteen devils captured in southern Tasmania, a heavily farmed part of the state. Only two had eaten wool. Yet *Australian Wild Life*, in commending the Bauer article, and quoting selectively from it, did not mention its nemesis Guiler and his central role in making the Bauer case possible.

Bauer's empirical observations are many, including those derived from a long vigil he kept observing a devil lair. He took photos, which he called the first ever of the animal in its native habitat. Accompanying one of these is a caption which is as good a description as any of the devil: 'Note his long nose, tough body and sturdy legs. He's made for travelling rough in scrub terrain'.[7]

Jack Bauer's 'Protection That Doesn't Protect', an important historical document, is reproduced here in full:

> This is a unique animal, a sort of living link between high marsupials and the most primitive of all. Once it lived on the mainland from which it seems to have vanished. Today it is making its last stand, its last fight for survival, in Tasmania, and in all the world there is only 26 215 square miles left to it.
>
> But even in such a small part of the world as this the Tasmanian devil, *Sarcophilus*, is encountering an ignorance which could exterminate it. Zoologists and government experts are unanimous that the devil fully justifies the Protected Animal tag it wears but for the last two years

many Tasmanian farmers and graziers have claimed it to be a ferocious poultry and lamb killer. 'Give us permits to kill him in traps or with poison' they clamor. Recently some farmers gave the press this statement, which embodies a kind of death sentence for the devil.

They wrote: 'The devil has been attacking lambs and poultry. Wombats are also disappearing. Unless something is done about the devil now, it could become as much a menace as the dingo is on the mainland.'

These gentlemen apparently cannot tell the difference between a devil and a dingo. The devil is just about the size of a terrier. The dingo is a non-marsupial and consequently has a much higher IQ than the devil. The rate of mortality at birth must be much greater in the marsupial devil whose birth is a harder and more precarious one than that of the dingo. It was probably the dingo that contrived to exterminate or deplete the devil family in the mainland. There are about 200 dingoes to every devil.

The farmers' statement was followed by another one from Dr Guiler of the Zoology Department, University of Tasmania. He said: 'The devil is rarely found in the southern regions. But recently 18 devils were captured and sent to me at the university and examination of their stomach contents showed that only two of them had eaten wool.'

Thus it appears that the devil will occasionally take a lamb but in the country in which it lives it finds it easier to take smaller marsupials, wombats, bettongs, potorus and the extremely prolific Thylogales or scrub wallabies. However some thoughtless farmers and graziers are killing devils on their properties. Yet how many lambs do they lose to the devils! How do they know that the few lambs that they lose are actually killed by devils? Unless they see devils in the act of killing their livestock they cannot be sure that these animals are the killers.

Tracks don't mean a thing in the rough, scrubby and rocky bush in which these graziers' sheep roam. As to poultry kills, these can be blamed on domestic cats gone wild which are very plentiful in most parts of Tasmania. Likewise, the native cats may be the killers. And wedge-tail eagles, crows and hawks may account for some lambs too.

There is very little known about the animal. The only reliable information on this animal is given by Troughton in his classic *Furred Animals of Australia.* He gives 58 lines on this animal. Not much to go by, but not much for farmers and graziers to pin killings on the devil.

The first record of its existence comes from Deputy Surveyor Robert Harris who in 1808 wrote: 'These animals were very common . . .'

It was probably its devilish appearance that earned it the present monicker. The adjective 'Tasmanian' was added when it was found that it lived only in Tasmania.

But does it really? Here is something that will put doubts in your mind. Mr Troughton writes: 'A devil was killed at Tooborac about sixty miles from Melbourne in 1912 which may possibly have escaped from the zoo or private captivity . . . Fossil remains have been discovered in other parts of Australia. The recent appearance of skulls found in various localities in Victoria, including portions found amongst bones of existing marsupials in a kitchen midden of the Aborigines supports the view that it may still exist there.'

Thus the 'Tasmanian' devil should be regarded as an Australian animal, an animal that may exist or not in the mainland but which is of the utmost importance to all nature lovers and outdoorsmen. It has been classified in the subfamily

Dasyurinae which includes all our marsupial predators, ie, native cats, tiger cats, the Tasmanian tiger (here again there is ample proof that this 'Tasmanian' animal lived in the mainland in ancient times from where it may have been exterminated by the more developed dingo when this was introduced here by Aborigines or Asian peoples many thousands of years ago) and the mysterious Striped-Marsupial Cat of North Queensland which appears to be no ghost but a reality.

These predators are in a way relations of the American possums which are also marsupial and carnivorous but are tree-dwellers whereas our Dasyurus, except for the tiger cat, are ground dwellers. Thus the devil along with the native cat and the Tasmanian tiger or Thylacine, *is the only ground-dwelling marsupial predator on earth. If it became extinct a whole animal species would disappear from the face of the earth.* Do you now realise its importance?

The thoughtless, meat-hungry early settlers of Tasmania contrived to exterminate the island's emu, almost blotted out the Forester kangaroo (it is wholly protected but its numbers are small) and managed to practically exterminate the Thylacine or Tasmanian tiger. They also poisoned, shot and trapped hundreds of devils before these animals became protected. Together with the settlers' slaughter a strange disease which has not been explained, seems to have attacked the members of the Dasyurinae in Tasmania some 50 years ago.

[All] Of a sudden, the outback people and the naturalists of the day realised that these animals had become rare and that some—like the tiger—seemed to have disappeared. What happened to them? Nobody knows. And that's why scientists are most concerned with the fate of the devil. For some reasons these primitive marsupials seem to have it harder to survive these days. And the devils have been able to survive in

Tasmania possibly because there are no dingoes here. Nor are there any foxes. Thus the devils—and the Thylacines—have no competition.

In four years in Tasmania I have encountered exactly 11 devils in the bush and have spent many hours observing three in captivity. From my observations I have concluded that it is almost impossible to establish exactly what is devil habitat. I have encountered devils in the rainforests of the Cradle Mountain Lake St Clair Reserve in the West Coast of the island. I have seen one on the tea-tree and gum bush along the Crotty Track also in the West Coast This part of the island has a very heavy rainfall. But I have seen devils around the drier, scrubby, rocky bush around Table Mountain, Central Tasmania and also in the high, bleak country around the Arthur Lakes.

Thus it looks like the devil has a wide range in the island, but I feel that his favorite habitat must be country of dense bush. He doesn't look like a 'savannah animal', a runner. He isn't a fast runner. Dogs can easily overtake him.

He is essentially a scrub animal. Physically he seems just made for the scrub country. His compact, tough body, his short legs are assets for him to run through brush and to crawl under logs and into hollow trees. As far as I know he isn't a tree-climber. But I do know that he is a very silent mover, another asset for him to hunt in scrub country and to approach his quarry silently.

It is always jet black with some spots or stripes on different parts of the body. Some I have seen had only one stripe across the throat, others had one or two spots on different parts of the body. Mainly, he is all black, though. And his color is an excellent camouflage in the scrub. Placing a black rag amongst some fallen logs I have failed to spot it from a distance. This

In the 1960s naturalist Jack Bauer spent long periods in the Tasmanian bush observing the devil. One night encounter took place near Damper Inn on the Port Davey track in far south-west Tasmania. (Photograph of Geoff Hood by Leo Luckman, Hobart Walking Club, 1940. Courtesy TMAG.)

same black rag could hardly be seen in the deep shadows of rainforests. It was also almost invisible from a distance in a stand of tea-tree gum and dogwood. Moreover, at night the devil's black coat must render him almost invisible.

They will tell you he is a 'nocturnal animal' but strangely enough of the 11 I have seen in their wild habitat only one was out at night, at 2.20 am near Damper Inn,[8] on the Port Davey track. If the devil preys on bettongs, wombats, potorus, scrub wallabies, etc, he must go hunting toward evening and in the early morning which are the times when these small marsupials are at their feeding or watering places.

Whether or not the devil and his quarry continue to roam about the bush all night through I don't know. I have spent many nights sitting it out in the bush with a spotlight in order to spot game. All I have seen were possums, bandicoots, an occasional scrub wallaby. Except for that one devil I have mentioned.

Observing three captive devils I noted that they spend a great part of the day slumbering in their cage. These devils are a mother and her two young, a 'gal' and a 'boy'. During the day they *always* sleep bunched up together or even one on top of another but never separated. Toward evening they wake up and begin to prowl around the cage snarling and growling. Each is a prime animal and eats one rabbit a day . . .

Devils don't make good pets, but they aren't dangerous animals either. The friend of mine who keeps three in captivity has never been able to chum up with the little devils. He can handle the females. They'll never attempt to bite him. But they don't show any affection either. But he is quite unable to put a hand on the male. He's a real tough guy. As soon as my friend enters the cage, the devil snarls and shows his fangs. But he is a small animal weighing at most 12 lb (this is the weight of a prime specimen) and he can be put out of circulation with a well-placed kick if he decides to attack.

Devils smell. I have struck two devil lairs (just a scrub-bed under a log) and the smell coming from them was appalling. Like all the other Dasyurus, the female devil's pouch is oriented backwards, that is, contrarywise to the kangaroo's. She seems to have a usual litter of two young which she carries in the pouch for perhaps three months or so. In the pouch there are two pairs of teats.

In the old days, devils were killed by dogs, shot by hunters, poisoned, and trapped. As you can imagine, shooting a devil is no great shakes as a sport but you do have to shoot quickly because although he is a slow runner, he can move plenty fast in the scrub. He moves in a lop-sided, shambling gait reminiscent of a bear.

In the kangaroo-snaring boom, some years ago, many devils were caught in the snares but they didn't stay caught for long. They chewed up the string and got themselves free again. They couldn't, of course, get away from a dingo trap, in which many were caught.

To see devils in their native habitat is not easy. Much depends on luck but a great deal depends on bushcraft also. First off, watch for droppings. A devil's droppings are about the size of a possum's but when crushed the devil's always contains some bones, feathers or hair, the result of eating flesh, whereas the herbivorous possum's dung contains only vegetable matter. The droppings of native and tiger cats are smaller. However, you may find droppings and still be far from finding a devil. How far does he range? I don't know. He may be here today and ten miles away tomorrow.

Hunting kangaroos with dogs is perhaps the easiest way to see devils. The dogs often flush them out of their hiding places. That's just what happened a few weeks ago when our dogs roused a devil from under a log in Central Tasmania. We let the little chap get away then I went to inspect the log. Just under it and completely hidden from sight, was his lair, a scrub bed badly matted with droppings and badly stinking. Nobody had ever taken pictures of a wild devil in his native habitat and here I had the chance.

A week later I was back there. I chose a hiding place about 250 yards from the devil's lair and watched it with a pair of binoculars. About 6 o'clock in the evening just at sunset, I spotted the animal emerging from his 'home' and hitting the hunting trail. I spotted him again at dawn the next morning. That afternoon—my second day there—I made my first preparations for pictures.

With the wind blowing against my face (just the way I wanted to make sure that he was still in those parts, the devil's nose) I approached cautiously his lair and began building a hide. I was cutting a pole with my jungle-knife when I heard a noise and looking saw the devil legging it away from those parts. He'd heard or scented me and decided to beat it. That was okay with me. I built my hide from natural vegetation about 20ft away from the devil's lair under the log. Then I shouldered my pack and made tracks for home.

A week later I went back there. By this time I hoped that the devil had recovered from the scare I gave him and had become familiar with my hide as well. However, I wanted to make sure that he was still in those parts. So that evening found me in a hiding place with my binoculars. Sure enough, roundabout sunset the animal emerged from the log and hit the hunting trail. As soon as he'd gone I took my camera with tele-lens and tripod, a thermos of coffee and some sandwiches and sneaked into my hide.

I had no intention of photographing him with flashlight because I was sure I'd spook him away from those parts. So I made myself comfortable in the hide and went to sleep. I am sure I heard him return to his lair some time in the night but it might have been the noise of a roo or of the wind I heard instead. The wind, by the way, was blowing hard into my face.

It was exactly 6.10 am by my watch when my heart almost skipped a beat as I crouched over my camera tripod inside the hide. I could just see the devil's head peering over his log. I made a telephoto picture. The click of the release button didn't seem to spook the animal. Perhaps he hadn't heard it. The wind was making plenty of noise.

Gingerly he clambered over the log and for a few minutes moved restlessly about the low scrub, his sharp-pointed nose sniffing the air on which he perhaps detected some suspicious smell. Mine. But the wind was being a great help to me. I made 17 pictures of the animal showing his various reactions before he finally hit the hunting trail, moving away in that bear-like gait that makes him look so clumsy and slow. So I had obtained the first pictures ever taken of a devil in his native habitat.

From my observations it looks like that animal—at least when I was there—left his lair at dusk, returned some time during the night and went hunting at dawn again. He probably returned shortly after sunrise. These times coincide with the feeding times of most ground marsupials. However, I should have had to spend several more days in that spot to gather more accurate information. But I couldn't spare any more days.

Let me hope that my great-grandchildren will also be able to see and to photograph this fascinating little animal, which represents one solitary species living on this earth. More than any other member of our fauna does the devil need to be wholly protected. So let us hope that the Tasmanian wild life authorities will not allow farmers to destroy these animals and let us also hope that all sportsmen and outdoorsmen who know of farmers who kill devils will report them to the game inspectors or police officers. A person killing a devil can be fined up to £100 [stg] and that is really a cheap price to pay for this offence.[9]

10

FROM ANTICHRIST TO AMBASSADOR

> In the late 1960s years we saw very few Tasmanian devils at the shack but then their population increased greatly . . . We were frequently accosted as we went out to the toilet. The children used to be quite scared of them and would come running back inside for an adult to accompany them . . . The devils became bolder and would remove items from the back porch: on one occasion they took a box of six-inch nails and we followed the trail of nails down to the beach where they had dumped the box.
>
> JENNY NURSE, HOWRAH

In the early 1960s there weren't many Jack Bauers, or Eric Guilers. It took the stubbornness and certainty of their kind to recast the Tasmanian devil, as Mary Roberts had half a century earlier, as an animal quite unlike the popular perception of it. At Granville Harbour Guiler and his colleagues worked in tough conditions, camping in rough terrain for up to ten days, setting baited, drop-door wire cage traps in rainforest, gullies, cleared farm paddocks, coastal scrub and dune formations. The rainy, soggy, windy, cold locality had been chosen for its inaccessibility and lack of human activity; just three people lived at the harbour, with the nearest township, Zeehan, a day's drive away on a rough track.

Among Guiler's findings over the ten years of the survey: devils have a home range but do not defend territory; they use well-defined tracks and livestock trails; they travel extensively in search of food; despite the adults being solitary, they may develop some form of social intercourse at a 'general mixing area';[1] few live beyond six years; a population may fluctuate rapidly and substantially, linked to both high juvenile mortality rates and the degree of immigration of animals into an area; devils in the west are smaller than elsewhere.

A sharp, sustained increase in numbers was recorded in the latter years of the survey, beginning 'very substantially' in 1973, with the boosted population showing 'a good balance between the old, mature, and juvenile weight groups'.[2] That balance had disappeared by 1975, and many animals, when captured in 1975, weighed less. Guiler speculated that this was due to a food shortage, although 'there was no field evidence that this was in fact so'.[3] An increased population in one area would surely mean less food for all, and animals would soon lose condition, negatively affecting reproduction ability—a possible mechanism for self-correcting population imbalances.

Anecdotal population evidence, in the form of newspaper accounts, is one source of information, but it is patchy, inconclusive, and open to interpretation.

In the winter of 1966, possum hunters across a small area of the midlands reported great increases in devils robbing their snares. This was put down to a rise in their numbers. The devils 'appeared to be very ravenous, according to hunters'.[4]

That behaviour suggests either a lack of food, or more vigorous competition through an increased devil population, a natural winter result of weaned juveniles. To muddy the picture further, possums were apparently more plentiful than usual in the eastern area but scarce in the western area. What, then, caused an apparently isolated outbreak of atypical behaviour?

In 1972 the same area appeared to be afflicted again, though this time farmers were the victims. 'It is believed that the animals have increased so much in the timbered country that they are venturing into the open lowlands in search of food. Some farmers are becoming concerned that they could eventually attack stock.'[5] 'Evidence' for this included the discovery of a devil hiding behind a deep-freeze unit in a garage in Oatlands.

A dramatic report then declared:

> Devils loom as menace . . . Tasmanian devils had recently attacked chained farm dogs, a spokesman for the Tasmanian Farmers, Stockowners and Orchardists' Association [TFSOA] claimed yesterday. They had savaged domestic animals and had been found inside at least two farm houses. 'The build-up in the population of this dangerous pest is alarming', the president of the TFSOA (Mr R. J. Downie) said. 'We are getting reports which strongly suggest a population explosion among devils. They are being reported from places where previously devils were unknown, or at least, not a problem. They are readily attacking stock, prowling around and in farm buildings, and fighting with chained farm dogs,' Mr Downie said.[6]

The president went on to say that his association recognised the role of native fauna and would never advocate their mass destruction; but there is little doubt that his awareness of the devil being something of its own advocate—a threat while a boon—came from Dr Guiler, who had not long before warned farmers that, 'a permit [to destroy devils] is the last line of defence. It is a recognition of failure to keep pastures clean and free from dead animals. A buildup of devils to plague proportions can only occur if food is available. In this respect the devil takes advantage of man's untidy habits'.[7]

The difficulty with such accounts is in knowing what to believe. An island-wide increase would surely have been commented on elsewhere. Yet what conditions might create a purely 'local' increase? Food supply—including dead and vulnerable sheep and cows—plays a part.

For fifteen years no more was heard of the devil as a problem, but in 1987 another population surge made it newsworthy again. Launceston's daily newspaper the *Examiner* kicked off:

> Farmers down Cranbrook way are having a devil of a time, and would like the power to legally do something about it. The small east coast district is beset with Tasmanian devils, but as a wholly protected animal the ferocious little marsupials can maraud hen houses with the full protection of the law . . . A farmer's first concern is for his stock, so when someone or something is crunching his critters his instinct is to shoot first and ask questions afterwards . . . Although local folk agreed that an unusually and annoyingly large number of devils had been roaming the district lately, most thought the word 'plague' was a bit too strong. 'They're a pest, not a plague', said Jim Amos, of Cranbrook House, who has lived and farmed there all his 70-odd years. A story in yesterday's *Melbourne Sun* told how 'a plague of voracious Tasmanian devils is causing havoc on the east coast' . . . Merino stud farmer

> Geoff Lyne claimed to have lost 75 ducks and 50 hens to the little devils in the last few months, [they] had attacked ewes during lambing and savaged a $2000 ram as it lay trapped in an irrigation ditch . . . Most farmers and townsfolk had lost chooks to the mean marsupials, but Mr Amos said that devil numbers had been up for 15 years, and he had even had them living under his house. 'They're like possums now—they're dead on the road all the way to Swansea.' Mrs Ethel Poole, 72, of Cranbrook, said that devils had carried off all but two of her chooks. 'They're thick as fleas around here, and they're sly things—they left me with two old boilers,' she laughed.[8]

In June 1987 the rural *Tasmanian Country* took a less light-hearted view:

> Farmers in Tasmania's North-East are concerned about large numbers of Tasmanian devils in the area. While the devils are causing problems with livestock, the farmers are at a loss to know what steps can be taken about what they consider to be plague proportions of the wholly protected native animals . . . Waterhouse farmers, Lindsay and Lois Hall, say they can barely set foot outside their back door without running into devils. Mrs Hall said that 25 years ago she would see 'the odd one'. However, she said, on their cattle grazing property, large numbers were now seen during the daytime and were creating havoc. Mrs Hall said the Tasmanian devils took their chickens and ducks and chewed the ears and tails off newborn calves if they were too slow to stand up. She said the devils had also been known to take a litter of pups.
>
> Mrs Hall said she believed the problem started when, about six years ago, large numbers of Tasmanian devils were brought to the area from Cape Portland where they were becoming a nuisance. Since then they had bred up to plague proportions.

> Mrs Hall believes because there were so many Tasmanian devils in the area there was a shortage of food and the animals were in poor condition and mangy. 'I wouldn't like to see them go because they are a unique animal,' she said.[9]

Mrs Hall's sympathy for the devil was admirable, given the dislike of it held by many in rural industries. But a few others were beginning to see a different value in the animal. A metamorphosis, dollar-inspired though it may have been, had begun back in the mid-seventies when a prominent businessman and member of the then Tasmanian Tourist Authority suggested that the devil be used to attract tourists. A sixteen-week global study trip had shown him:

> Tasmania is almost unheard of throughout the world, and those who know of our State, know it only because of its connection with the Tasmanian devil . . . Now is the time for us to sell, sell, sell our natural product through our tourist industry . . . Many

The official logo of the Tasmania Parks & Wildlife Service. (Used with permission of and © the Tasmania Parks & Wildlife Service)

> parts of the world are suffering from the thoughtlessness that has accompanied industrial expansion, and we must do everything we can to preserve our unspoiled environment.[10]

Luring tourists to a place for its wildlife is understandable. Taking an animal out of its natural habitat to attract tourists to that place is a different matter. So it was that in 1981 the Tasmanian government proposed to use a devil as a central feature of a tourism task force visiting New Zealand. The animal would then be donated to a university for scientific study. It's hard to imagine a lone devil in a cage enticing anyone to visit its homeland, and the Australian government in any event refused an export license. When this failed, donation became the next option, and attention turned to Japan.

The State government found a way around the issue, donating four devils to Osaka Zoo in 1984. According to the official ministerial news release, they were being given

> to the Japanese people . . . The gift of the devils will help to cement the bonds which are being developed between Tasmania and Japan . . . The devils will be unusual but important ambassadors for our State . . . The Japanese people were fascinated by the devils, and their interest in the animals would focus attention on Tasmania. The Japanese press had already given extensive coverage to the pending arrival of the animals, describing them as the 'strangest of all animals' and 'with strong teeth, even to bend iron stick' . . . Osaka Zoo officials had prepared a special home for the devils, and plans also were being made for an official receiving ceremony.[11]

A few years later three more devils, the youthful Mo, Mavis and Mary, were presented to the Sapporo Maruyama Zoo in

Japan's Hokkaido state. The Tasmanian official accompanying them on the flight, Ray Groom, the Minister for Forests, Mines and Sea Fisheries, expressed the hope that they would breed in captivity, again 'putting Tassie in the spotlight'.[12] Needless to say, they didn't breed, but as a marketing ploy it worked; the devil was to become as significant as the koala as an iconic Australian image.

This new-found respect for the animal in its home state was not before time. While devils continued to be regarded as pests in some agricultural areas, public sensitivity to its status rebounded on the university, which since the time of Flynn had been associated with research for its protection. A saga which made international news in 1985 started with a front-page report in *The Mercury* headlined 'Uni's Devilish Experiments Anger Animal Libbers':

> Animal Liberationists have warned they will picket and possibly invade a University of Tasmania seminar in Hobart tomorrow to protest against experimentation on and slaughtering of at least 11 Tasmanian devils. The seminar, in the university's zoology department, has been arranged for the presentation of reports on research to ascertain the temperature regulation of Tasmanian devils' brains.
>
> The experiments were by an honours student working on his Bachelor of Science honours thesis. A spokesman for the Tasmanian chapter of Animal Liberation, Mrs Pam Clarke, yesterday said the experiments had been futile. Several animals which had had sensitive temperature recording instruments called thermocouples implanted in their brains had been found to be useless for the experiments because the thermocouples had corroded. 'The devils, a part of our unique wild fauna,

> have been through a horrendous series of experiments,' she said. 'We were horrified to read that many of them died during the implantation operations and also during other experiments,' she said.
>
> The survivors had been forced into prolonged exercise on an enclosed treadmill. An electric shock grid had been put at the rear of the treadmill 'to encourage the animals to continue running', but this was discontinued because 'it caused unnatural responses and also affected the chart recorder'.
>
> Mrs Clarke said the distressed devils had suffered substantial injuries to their tails and paws when caught between the treadmill and the boundary wall. Animal Liberation also has claimed that an unspecified number of native cats and possums have been slaughtered in university-sanctioned experiments. 'Animal Liberation calls on the university to open its doors on the secrecy surrounding animal experimentation and appoint a member of an animal welfare organisation to its animal ethics committee,' Mrs Clarke said.
>
> The head of the pathology department in the university's medical faculty, Prof Konrad Muller, yesterday defended the experiments on the grounds that the research was important. A similar appraisal of another animal's brain, for instance a rabbit's, would not have given the desired results. The 11 Tasmanian devils had been ordered by Dr S. C. Nicol, of the university's physiology department, with the permission of the State National Parks and Wildlife Service. Of the 11, six had been killed and their brains immediately examined. The other five had been used in a series of tests to determine the regulation of their brain temperatures.[13]

These allegations were not met with silence. The story continued on the front page of the next day's paper:

> A senior lecturer at the University of Tasmania yesterday lashed out at what he called ignorant and ill-informed criticism of experiments . . . they had considerable scientific merit and had resulted in the discovery of a blood cell which controlled the temperature of the marsupial's brain. 'The experiments have a number of implications to the evolution of marsupials and the evolution of mechanisms which keep body temperature constant in all mammals, and for understanding the devil's way of life,' Dr Nicol [sic] said . . . He said the results had been enthusiastically received by the *Australian Journal of Zoology* and another Australian university which was doing similar experiments . . . The experiments were part of a thesis by an honours student for his bachelor of science degree. Dr Nicol said the student was not Australian, and poor expression had made the experiments appear worse than they really were . . . Dr Nicol said that [the thermocouples] had broken in their rubber casing, and were useless for the experiment, but had not caused the animals any extra discomfort. He also dismissed claims that devils had been forced to run for unnatural periods of time on a treadmill. 'The treadmill experiments involved only two animals which ran at 7 kmh . . . In the wild the animals keep this sort of speed up for hours.'[14]

A ministerial statement defended the university, which was not surprising given that the government had issued the experiment permits in the first place. The minister curiously observed that devils were in abundance, as if that overrode questions of ethical treatment of individual animals. The issue duly blew over.

Devils weren't long out of the news, however. Tasmanians awoke one morning in July 1988 to a front-page horror headline: 'Devil's Disease—State's Tough Little Ambassador Threatens Livestock'.[15]

The discovery of the deadly animal parasite *Trichinella spiralis* for the first time in Australia—in devils—had potentially disastrous national ramifications. Not only might it migrate to livestock (pigs are the main host) or to humans (causing eye and heart damage), but both Tasmania's and the country's disease-free livestock status might also be seriously jeopardised. The infected devils all came from an area near iconic, isolated Cradle Mountain. How could a foreign parasite make its way there? Tourists? It was speculated that a devil or devils must have eaten an infected product, most likely illegally imported salami or biltong, because curing and smoking meat doesn't kill the worm.

Fortunately, the threat proved to be an unintentional beat-up. Sampling outside the Cradle Mountain area by Nick Mooney revealed the worm to be naturally present in about 30 per cent of the devil population. Government veterinary pathologist

Nick Mooney for many years had a key official role in the management and research of Tasmania's wildlife. (Kate Mooney)

Dr David Obendorf confirmed that no crossover risk existed. 'Wherever you find Tasmanian devils you find the parasite but that's no reason for killing Tasmanian devils.'[16] It was a strange way to discover more about the tough little ambassador. Further studies confirmed that *Trichinella pseudospiralis* occurs naturally in a wide range of mammals and birds, including some quoll and possum species.

In the space of twelve months, in 1991 and 1992 three very different accounts of the devil were published by the world's foremost devil experts. The University of Tasmania conferred David Pemberton's doctoral thesis, 'Social organisation and behaviour of the Tasmanian devil, *Sarcophilus harrisii*'; Eric Guiler published his 28-page *The Tasmanian Devil*, which covers a year in the life cycle of the animal; and Nick Mooney's 'The Devil You Know' appeared in the Winter 1992 edition of *Leatherwood: Tasmania's Journal of Discovery.* Despite its brevity Mooney's article is one of the first genuinely informed accounts of the animal written for the general public. Mooney had been studying and interacting with devils for years; the importance of 'The Devil You Know' is its wealth of previously unpublished and logical assumptions about the animal, including:

- The demise of the thylacine probably resulted in diminished competition for, and predation on, devils. It is also reasonable to suppose that the niche of devils then expanded as it has for hyaenas as the number of lions diminished in Africa. I wonder if devils now may be of a larger range size and more predacious than before, gradually evolving to soak up the empty (or good as empty) thylacine niche.
- It is a pity that the first exotic eutherians our marsupials had to deal with were probably the very cream of that group as

far as survival goes: humans, dogs, foxes, cats and rodents. For a long time this 'unfair' competition has clouded the true success of marsupialism.

- Small devils have a variety of natural competitors and predators including (previously) thylacine, people, other devils, quolls and large birds of prey. Eagles and people are probably two of the main reasons devils and many other Australian animals are nocturnal, directly to avoid predation and indirectly to minimise competition.
- Unusual items I have found in devil scats include: part of a woollen sock; a wallaby foot complete with snare; part of a dog or cat collar; 27 whole echidna quills; stock ear tags and rubber lamb 'docking' rings; head of a tiger snake; aluminium foil, plastic and Styrofoam; ring off a bird's leg; half a pencil; leather jacket (fish) spine; boobook owl foot; cigarette butt; part of a 'steelo' pot scraper. I have also had part of a leather boot and the knee of a pair of fat-stained jeans eaten after being left outside a tent (not with me in them).
- I have made some observations of sheep and lamb–devil interactions using military style 'starlight scopes'. Large devils will check out a flock by sniffing from 10–15 m. The sheep will group and face the devil, stamping their feet as their usual threat. If the sheep are all healthy and alert and no carrion or afterbirth is available the devil(s) quickly move on. Sick or injured stock attract much more attention. Healthy sheep without lambs usually ignore devils.
- Although devils use their extraordinary strength to escape traps they rarely use it to enter places to eat.
- The mechanism of foraging seems to be almost ceaseless patrolling . . . I have followed individual devils for more than 11 km along beaches and through the snow before losing their tracks.

The devil's varied and indiscriminate diet results in disproportionately large scats. (Courtesy Nick Mooney)

- Human interference can be important, either by providing extra food or extra mortality, especially with illegal poisoning. Often, as in some rural areas, it is a bit of both resulting in unusually high population turnover.[17]

Like the earlier articles by Mary Roberts and Jack Bauer, Mooney's field observations cut right through much of the dogma that continued to be associated with the devil.

11

IN CAPTIVITY

> At the devil pen the rock wall prevented my son from catching a good view, so I picked him up and we leant over for a better view. My overpriced but much-loved sunglasses fell into the pen. I contemplated retrieving them, but we then watched in awe as two devil diners crunched silently on them until they had completely devoured them. Since then I've never spent more than twenty dollars on a pair of sunglasses.
>
> RICHARD PERRY, WEST HOBART

Tasmanian devils are easy to capture and keep captive in secure runs. Nocturnal by preference, devils in captivity are usually displayed during daylight and fed at optimal visiting times. They have always been regarded as curious creatures, a legacy of nineteenth-century attitudes ranking marsupials as inferior to placentals. As noted, they were displayed in zoos across the world from the mid-1800s until well into the twentieth century, when export restrictions came into place. This chapter looks at two contemporary instances of devils in captivity.

Angela Anderson was resident zoologist at the Tasmanian Devil Park, now called Unzoo, a tourist attraction and wildlife rehabilitation centre at Taranna, near the Port Arthur Historic Site on the Tasman Peninsula. Anderson studied in Glasgow before taking up an internship in wildlife rehabilitation at the Wildlife Centre in Virginia, USA. There she specialised in birds of prey, treating up to 40 at a time. Across the world at the Tasmanian Devil Park a new raptor rehabilitation centre opened in 2001 and Anderson successfully applied for the position.

The Tasmanian Devil Park at the time had a resident population of about fourteen devils. Their enclosures have sturdy metre-high walls to allow for easy viewing. However, this ease of viewing can present problems. Captive devils are inquisitive and will stand up against the inner wall, and visitors have been

A Bonorong Wildlife Sanctuary devil, cared for by Director Greg Irons and colleagues and admired by tens of thousands of visitors. (Courtesy Bonorong Wildlife Sanctuary)

known to try and pat them. Children are sometimes held over the walls by their parents for a closer look, despite bold signs warning of the dangerous bite of devils. None of this is advisable. The question most frequently asked by visitors is whether they can pick one up.

On one occasion a child's parents purchased a soft toy when entering the Park. The toy was a devil, and a real devil managed to get hold of it. The child was distraught. Anderson entered the enclosure, gave chase and retrieved the toy, which was only a bit damp. This is an aspect of the animal which puzzles US visitors in particular; their expectation is that, like Taz, it has a blinding turn of speed.

Spring and early summer visitors are likely to see unweaned pups. In January 2004 one of the enclosures contained four newly independent siblings, their mother having been returned a month before to the nearby adult enclosure, and another four still with their mother because of the persistence of a late suckler. The independent four continued to display a great sense of bonding, sleeping all over each other.

The two mothers were sisters, who three years earlier had arrived at the Park as roadkill pouch-young. The adult enclosure's patriarch was seven-year-old Max, whose final litter was approaching two years in age. The Park's multiple bloodlines permit a number of breeding combinations—a necessary captive breeding strategy.

Educating visitors as to the specific implications of DFTD, and of the extinction threats to so many other native species, is a significant aspect of the Park's role. Anderson, a great admirer of devils, noted that they are 'very shy. The devils we have were reared by their parents or brought in from the wild and we've never had any problems with them at all. The only devils that I have seen to be aggressive are the ones that were hand-reared. They've got no fear of humans'.[1] She could attest to that. Once she was in an enclosure bending down to get a water bowl when a devil that had been hand-reared latched on to her shin. She shook it off by leaping out of the enclosure but has a scar as a permanent reminder of the incident.

Foraging behaviour is encouraged through varying types and amounts of food. On one day a number of small pieces will be tossed into the enclosure; on another it may be just one large piece for them to wrestle over. Two or more devils tugging at a chunk of meat and bone may look like competitive feeding but it is in fact cooperative, because in this way the food is quickly broken into manageable pieces.

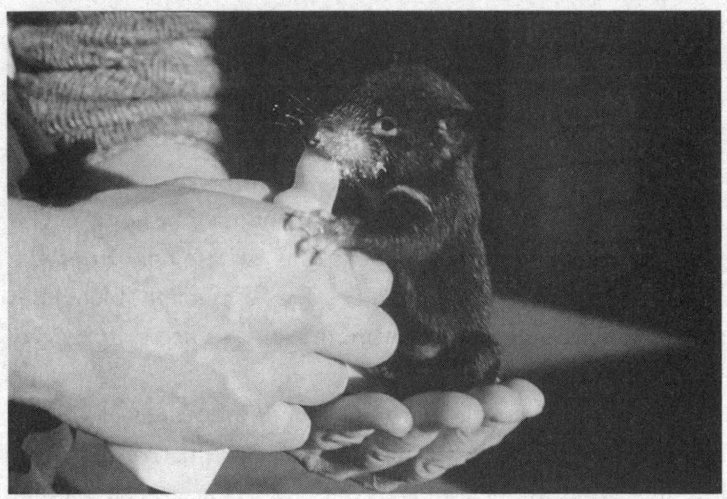

Live baby devils are frequently found in the pouches of their roadkilled or disease-killed mothers. They are kept in captivity by human carers (to whom they readily take) until old enough to fend for themselves in the wild. (Courtesy Nick Mooney)

An innovative, and energetic, feeding method is the suspension of meat on a bungee cord for the animals to leap at. Meat pieces are also hidden about the enclosures, in cardboard boxes and in toilet roll centres. Roadkill, generally commercially shot wallaby, makes up a large proportion of the food, along with laboratory rats, and rooster carcasses from chicken farms. Devils are partial to dog biscuits and eggs. Anderson has observed a mother attempting to encourage her young from the den by running up and down outside with food in her mouth. On the other hand, there are times when the mother will take food from her young.

Tasmanian devils still living in the wild on the Tasman Peninsula represent the last isolated, natural population of disease-free, wild devils in the world. Unzoo has a long history of supporting the effort to save the Tasmanian devil and is a partner in the official Tasmanian Devil Conservation Project. This critical project aims to save Tasmanian devils on the Tasman Peninsula by preventing the spread of DFTD to the region. As part of this effort, Unzoo is breeding healthy devils for future wild release on the Peninsula. Unzoo also maintains a special devil-proof barrier fence at Dunalley, designed to prevent the spread of DFTD into the disease-free Tasman Peninsula region. In addition, through its Devil Tracker Adventure project, staff constantly monitor local wild devils and collect important information on the local devil population through infra-red cameras and data recording.

Fort Wayne is a city in northern Indiana. Its story is in many ways a microcosm of US history. In 1794, after defeating the indigenous Miami peoples, General Anthony Wayne built a fort at the junction of two rivers in order to facilitate permanent white settlement. The arrival of the railroad in the 1850s significantly boosted the town. Today it's a manufacturing centre surrounded by fertile farmland, with a population of about 170 000—similar to Hobart's. Among its attractions is a children's zoo. The opening of its Australian Adventure in 1987 was the crowning event of that year. The zoo broke all attendance records, and received a prestigious award from the professional zoo community.[6]

A Tasmanian devil named Coolah became a star attraction. When Coolah died in May 2004, he was seven and a half years old: possibly the world's oldest devil. Elaine Kirchner, supervisor of the Australian Adventure, looked after Coolah throughout his life at the zoo. Her experiences were summarised in an email interview with David Owen.[7]

> *David Owen*: You've looked after twelve devils in seventeen years—you must have a very good understanding of the nature of the animal. Do they differ much in personality? Any interesting examples of personality contrasts?
>
> *Elaine Kirchner*: Tasmanian devils, like most animals, do have differences in 'personalities'. Some are very much loners and don't want anything to do with other animals. On the other hand, we had two sisters, Rosie and Kestra, that nearly always slept in the same nest box. Every few days they would move all of the nest material to a different nest box and set up housekeeping there. We generally housed our devils singly, but Rosie and Kestra were very close. The only time we mixed more than these two together was during breeding season. We have found that during most of the year the female devils are dominant. It seems that when breeding season rolls around, the males assert their importance, if only for these few days. Frequently the males will drag the females around the exhibit by the scruff of the neck. They often corner them in a nest box, not allowing them to leave for up to two or three days.

Some animals were very curious and routinely explored every corner of their exhibit each morning. Others were content to go for a run around the place and settle in to sleep. Many years ago we moved them to an outdoor area, and then added four shelter areas for them. These were triangular, covered with brush and wood chips. All faced the public but were at various distances from the boardwalk. Some animals chose to sleep in the one closest to the public, some far away, and often an animal slept behind the shelters!

New animals in the exhibit always seem to be a challenge. When we got our two sisters in 1996, we had our assistant volunteer watch the animals to make sure they stayed in the exhibit. After about an hour or so, we heard a frantic call from Dianna saying that one of the youngsters was at the top of the fence. Dianna was armed with only a radio and a clipboard. Needless to say, all of the animal care staff moved quickly to contain the devil and modifications were immediately made to the hotwire!

David Owen: The devil's reputation for being ferocious is unfounded. But you surely must have developed a good respect for their inquisitiveness which includes a preparedness to bite. Did you have any unpleasant encounters over the years? And if so did you develop any methods to counter that instinct? And your broader interaction with them?

Elaine Kirchner: We learned over the years to have a healthy respect for all of those teeth and those strong jaws. My first encounter with an uncontained devil came early one morning, only a few weeks after we had received our first two animals from Australia. We had been told that devils don't climb, and so had them housed in holding areas built of concrete block with wooden fences. The enclosure was just over a meter in height. These animals were less than a year old and quite agile. When I arrived at work one morning, I noticed that one of the animals was not in her pen. A quick look around told the story—she was on top of our bird holding cages, way back in the corner. I quickly called the assistant director for help, he crawled up on top of the cages with a net to recapture the critter. I was in the aisleway underneath him, along with a cage housing two angry water rats. The only flaw to this plan was an unsecured piece of wire. The animal went through the hole and into the hall with me. I had no net, and no way to get to one without letting the animal escape. We chased each other around the water rat cage several times until I managed to get a cage door open and the animal went into the cage. We later learned that the folks who told us that devils don't climb had one animal who was quite old and very obese!

In general, we don't enter the enclosure with the devils. If we must, we use a large warehouse broom to keep between ourselves and the animal. They seem not to know that they can go around the ends of the broom. When we need to restrain them, we try to grab them at the base of the tail. It takes several keepers to hold one down for blood draws. Since they have virtually no neck, someone needs to have rather large and strong hands to hold down the head.

David Owen: Your website mentions devils at San Diego, Cincinnati and Toronto zoos; also a figure of some sixty devils in North America since 1983. Did Fort Wayne begin with a consignment from Australia or did you obtain captive stock from other North American zoos?

Elaine Kirchner: Eight of our animals came from Australia, the rest from North American zoos.

David Owen: I hope Coolah's enjoying his later years. What can you tell us about him?

Elaine Kirchner: Coolah is a rather small animal as devils go. He is nearly all black, just has a smallish patch of white on his rump. His teeth seem to be in pretty good shape. We do give our devils some whole prey but most of it is in chunks. They eat slices of rabbit, whole day-old chicks, fish (though he always bites off the heads), dry dog food, and the occasional meat ball. His birthdate is listed as January 1, 1997 at the Cincinnati Zoo, though I don't know if that is an actual date or an estimate. In looking at the records I just noticed that his grandfather actually lived at our zoo also for a few years at the same time Coolah was here.

Our first Tasmanian devils arrived here in June of 1987, just at the time we opened our Australian area. In 1989 we acquired four more animals from Australia, giving us a total of six.

As an older animal, Coolah does have some health issues, but still continues to hold his own. Our zoo opens for the season in a few weeks, so he'll be back in his exhibit then. Until that time he has two indoor pens and a couple outdoors as well. Just yesterday he was laying in the sun in mid-afternoon, acting like he had not a care in the world. He looks forward to his morning chick—even if it has medicine inserted inside it. Most mornings he comes out of the box to greet me, though he does like to sleep in at times. Now that it is light [early] in the mornings, he most often gets up and looks for his morning treat, though that comes a bit later.

David Owen: Finally, did you need to have much contact with Australian wildlife people?

Elaine Kirchner: I do have contact with a number of Australian zoo folks. Androo Kelley from Trowunna Wildlife Park has been a great help with devils, and actually visited us several years ago. I have worked with a number of Australian

> species—echidnas, wombats, dingoes, Eastern grey kangaroos, Tammar wallabies, kowari, lizards, & numerous bird species. I have hand-reared two kangaroos and a wallaby. They lived with me for up to a year and went everywhere I went—I wore them in a pouch around my neck during the day, hung them on the doorknob at night. I am a co-studbook keeper for grey kangaroos in North America, and devote a lot of time to those wonderful creatures.

Elaine Kirchner contacted David Owen not long after the conclusion of the interview, on 23 April 2004:

> I need to let you know that we have unfortunately discovered that Coolah has inoperable cancer. He has a large tumor on his backside that is also infected. We will still allow him to be on display and will probably have a sign detailing his health problems. I see no reason to banish him from his exhibit just because he is ill. I will let you know how things are going with him—right now he in on an antibiotic to clear up some of the infection and some pain medication. We have done an ultrasound and brought in a couple of consulting vets.

Her next email read:

> I'm sorry to have to inform you that we had to euthanize our Tasmanian devil Coolah this morning. He had an inoperable malignant tumor on his backside. We've let him enjoy the sunshine and be on exhibit when he felt like it for the past few weeks. In the past week he has stopped eating—never a good sign for a Tassie devil.[8]

12

'THE SPINNING ANIMAL FROM TASMANIA'

I was lab manager in the Zoology Department at the University of Tasmania from 1962 until 1979 and I worked closely with Dr Eric Guiler. On one occasion he had at the department a number of devils in cages built into a room. Three in all, separated by steel mesh and right to the ceiling. There was one light bulb over the centre cage. Sometime during the first night the devil in that cage climbed up the mesh and chewed off the bulb and its holder. All that was left were two small pieces of electric cable sticking through a small hole in the ceiling. A good thing the light had not been left on.

RUSSELL WHEELDON, SANDY BAY

The story of how a small nocturnal marsupial carnivore came to be immortalised as a Hollywood cartoon icon is an unlikely one. It involved chance, luck, unsolved intrigue and a clutch of dramatically different personalities, among them movie mogul Jack Warner, Hollywood artist Robert McKimson, film star Errol Flynn and his father Theodore.

In 1883 a young Russian Jew, escaping the threat of Tsarist pogroms, arrived in New York. His surname may have been Varna; immigration authorities anglicised it as Warner. His wife later joined him in Baltimore where he had opened a shoe repair shop. Some ten years later their young sons, Harry, Sam, Abe and Jack, lured by the thrill and potential of nickelodeons, pooled together to buy a broken Kinetoscope projector (the silent film *The Great Train Robbery* came with it). They repaired the projector and screened the film in a tent in their backyard. So began the illustrious cinematic career of the Warner brothers.

Jack, the future all-powerful head of Warner Bros., had to wait until he turned sixteen before his brothers allowed him to become a formal partner in their grandly named Duquesne Amusement Supply Company. That was in 1909—a notable Tasmanian devil year. Thousands of kilometres west, at 42° South, Theodore Thomson Flynn and his pregnant wife Lily had just arrived in Hobart from Sydney, where he had been appointed the Ralston Professor of Biology at the University of Tasmania. (As a point of comparison, Flynn's annual salary was A$500; the Warner brothers were together pulling in US$2500 per week—well over five hundred times the professor's earning capacity.)

Theodore Flynn's enthusiasm for Tasmania, which was not matched by Lily's (she disliked the cold and missed her Sydney friends), soon led him into original terrestrial and marine research, some of the most important of which involved the Tasmanian devil. And Flynn is credited with being one of the first scientists to warn of the thylacine's impending extinction.

In June 1909 Lily, soon to rename herself Marelle, gave birth to their first child, Errol. Much has been written about famous

Hollywood movie star Errol Flynn's relatively short, tempestuous life. The oeuvre constitutes a mass of contradictions. Even his autobiography *My Wicked, Wicked Ways* (with its devilishly named opening chapter) doesn't exactly set the record straight, disarming and uncomfortably honest though it generally is when he's not cracking jokes, many of them cheerfully libidinous. Throughout the book, however, Errol Flynn writes respectfully of his father. He admired Theodore's intellect and achievements in biology,[1] just as years later in Hollywood he admired, when not loathing, his boss Jack Warner.

At the university, Professor Flynn's duties were divided between lecturing, examination and research, part of which was a requirement to research the diseases of plants and animals. He made no mention of disease affecting any dasyurid. Within a year he had completed one of his more important papers, which today is still regarded as a standard Tasmanian devil reference text.[2]

Ranked somewhere between low soap opera and high intellectual and artistic achievement, the Flynn family saga is a compelling one, though father and son tend to be treated as different species in the literature, while Marelle is variously described as vivacious, fun-loving, cruel to Errol and incompatible with Theodore; Flynn was:

> a tall, handsome man, patient with Errol, overfond of alcohol, somewhat shabby for a distinguished professor [and] as a contrast to his wife, so full of life and gaiety, Professor Flynn was often moody and looked ill at ease in the company of others . . . wishing that he was back at his home or at the University laboratory surrounded by his beloved animal specimens.[3]

Despite personal difficulties, including separation from Marelle, Theo went on to a career of considerable personal achievement. In 1930 he left for Belfast where until retirement he held the Chair of Zoology at Queen's University and became a member of a number of eminent societies, a far cry from the early years of bringing up naughty Errol—including the occasion Theo found himself in trouble with the Tasmanian Museum and Art Gallery from which he had borrowed skeletons of a devil, thylacines and platypuses for research purposes and not only not returned them for years but Errol had apparently damaged them.

All the while Warner Bros. Pictures Inc. was benefiting from utilising animals. Soon after setting themselves up in Hollywood the brothers began producing short serials using tame animals from a nearby zoo, in which a heroine would be 'chased' by a doddery old lion, tiger or gorilla and the serial suspended at a climactic moment until the following week. Then, in 1923, Jack Warner had the prescience—or luck—to take on a script in which a dog rescued a Canadian fur trapper (*Where the North Begins*). The search for a canine actor uncovered Rin Tin Tin, a highly trained German shepherd. The dog became Warners' first superstar, earning millions in a seven-year career. After Rin Tin Tin's death Jack kept up the animal flavour by introducing a horse, Duke, and its faithful owner, a young John Wayne. It is not surprising that Warner Bros. then took to the animated cartoon business with such gusto, since the brothers knew how positively audiences reacted to animals.

Errol Flynn refers a few times to the Tasmanian devil in his autobiography, including this (his father's?) definition: 'A Tasmanian devil (*Sarcophilus ursinus)* is a carnivorous marsupial known for its extreme ferocity'. Errol had a deep interest in the natural world from early boyhood. He loved the sea and its creatures and much of this came from Theodore, who also kept marsupials at home for research purposes. They were pleasures in a place of friction: according to Errol his mother found him unmanageable, and

> a devil in boy's clothing . . . My young, beautiful, impatient mother, with the itch to live—perhaps too much like my own—was a tempest about my ears, as I about hers. Our war deepened so that a time came when it was a matter of indifference to me whether I saw her or not . . . The *rapport* was with my father . . . When school finished, I raced home to be at his side, to hurry out into the back yard, where we had cages of specimens of rare animals. That courtyard was a fascinating place for a small boy. Tasmania is the only spot in the world where three prehistoric animals, the Tasmanian tiger, the Tasmanian devil and the animal Zyurus are found. Father had specimens of all of these in his cages, as well as kangaroo rats, opossums, sheep. I got to know these creatures very well, even the most savage, and I hated it when he had to chloroform one and dissect it . . . Occasionally I went with him on a trip in quest of one of the rare Tasmanian animals. We headed for the western coast, a difficult terrain, where there were huge fossilised trees. We hunted the Tasmanian tiger, an animal so rare it took Father four years to trap one.[4]

Errol alone knew what a Zyurus was, though he may have been relying on memory. His father had made a major palaeontological discovery in Wynyard in northwest Tasmania of the oldest known marsupial fossil, *Wynyardia bassinia*. At a nearby site was the fossil *Zygomaturus*, a wombat-like member of the megafauna. At the site Flynn also discovered a new squalodont whale species.

After a period of adventuring as a young man in Australia and beyond, Errol acted in a cheap movie in England (his second) which came to the attention of Jack Warner. According to Flynn, 'Warner saw me popping around on the screen with a lot of energy.'[5] According to Jack, 'I knew we had grabbed the brass ring in our thousand-to-one-shot spin with Flynn. When you see a meteor stab the sky, or a bomb explode, or a fire sweep across a dry hillside, the picture is vivid and remains alive in your mind. So it was with Errol Flynn.'[6] The year was 1935: a wild, virile, dashing, swashbuckling Tasmanian devil had arrived in Hollywood.

Jack teamed him with the equally unknown Olivia de Havilland in *Captain Blood.* The movie made him instantly famous. Yet despite the rewards for the disgruntled, rebellious Hobart youth who'd struck the Hollywood jackpot, Errol Flynn came to begrudge as much as appreciate his luck:

> You were assumed to be Irish, your name being Flynn . . . Nobody knew or cared that my whole life was spent in Tasmania, Australia, New Guinea, England . . . Nobody believed me when I talked of that background. They didn't want to hear of it. They wanted me to be Flynn of Ireland.[7]

Still, he went on to make over 50 films, mostly with Warner Bros., until his death from a heart attack in 1959. That output of over two films a year, mostly in lead roles, is considerable, while he also found the time and energy to become the era's most colourful and controversial Hollywood identity. But he never lost his love for the sea, in particular, nor for animals. He sometimes arranged whale-watching cruises, one eminent guest being Professor Hubbs of the Scripps Oceanic Institute. He bred

Inspired by his father Theodore, Tasmanian-born actor Errol Flynn developed a lifelong devotion to animals. In the United States he was the first to breed lion hounds, also known as Rhodesian ridgebacks. The first chapter of Errol's biography is entitled 'Tasmanian Devil, 1909–1927'. (Courtesy Steve and Genene Randell, Errol Flynn Society of Tasmania, www.geocities.com/errolflynn1909)

champion lionhounds, a breed more familiarly known today as Rhodesian ridgebacks.

Flynn acrimoniously parted company with Warner Bros. in 1952, after 'a violent argument with Jack Warner . . . although we laugh at it today'.[8] And no doubt they did. In his autobiography Flynn claimed that he was one of the very few able to saunter into Warner's office and expect to be treated as an equal. Given Flynn's dominating and uncompromising personality, the outwardly gregarious Jack must have been pretty formidable himself. According to his son Jack Jr:

> [He was] the most complex and confounding of all the brothers. For years I have tried to find the keys to the labyrinth of my father's mind, but it remains now what it was throughout most of his lifetime: boxes within boxes, rooms

> without doors, questions without answers, jokes without points, scenarios based on contradictions, omissions, and deceit. His was the anguished story of a man driven by fear, ambition and the quest for absolute power and control . . . [9]

It's a harsh sketch. As harsh as those caricaturing Flynn as a rapist-paedophile-Nazi. How would two such apparent demons get on in a closed room? Even Hollywood might struggle to script it . . .

Flynn's subsequent battles with drugs and legal troubles overshadowed a successful quarter-century of moviemaking and it's hard not to imagine a degree of sympathy for him from his former employer, which he had served very well. Less than two years later, in 1954 (as Flynn, in Jamaica, steadily lost his looks and highly conditioned physique), Warner Bros.' *Looney Tunes* produced a feisty, energised, crazy, ravenous, fearless, wild cartoon character, unlike any yet seen: the Tasmanian Devil.

How and why did this animated marsupial come about? There is no documentation, or official confirmation, from Warner Bros. or anyone else, linking Errol Flynn and Taz. Flynn himself makes no reference to the cartoon character in his autobiography, but just three Taz cartoons had been made when he wrote his book, and he may not even have known of their existence—back then they were a combined eighteen minutes of obscurity.

In a little under 30 years of cinema cartoon art, a host of major animated animal characters had come to life. Aimed at children (despite plenty of adult wit and sophistication), the Warner Bros. animal stars were familiar and non-threatening: Bugs Bunny, Daffy Duck, Porky Pig, Sylvester the cat and Tweety the canary, Michigan J. Frog, Speedy Gonzales the Mexican mouse, and Foghorn Leghorn the rooster. The men of Termite Terrace (as the animation building was known) did break out of the domestic/farm mould with their skunk Pepé Le Pew in 1945 and then the ultimate chase-and-outwit duo the Road Runner and Wile E. Coyote (prompted by a wonderful personification of the coyote by Mark Twain) in 1949. All three were the creation of legendary animator Chuck Jones, already famed for Bugs Bunny.[10]

Jones and his colleagues Friz Freleng, Tex Avery, Bob Clampett and Robert McKimson were responsible for much of the 'controlled lunacy' of the *Looney Tunes* and *Merrie Melodies* output across those three decades.[11] McKimson created the Tasmanian devil (it was not then called Taz) in 1954. Why? North America has plenty of interesting and unusual wild animals. Furthermore, the real Tasmanian devil has no recognisable 'personality' and back then there was no antipodean Mark Twain to give it one—unless, of course, Errol Flynn did, through the legacy of his own dynamic, destructive, insatiable ways (three adjectives which closely fit Taz). On the other hand, McKimson had created a marsupial six years earlier, Hippety Hopper the baby kangaroo. Unlike the then-obscure devil, the kangaroo had long and famously symbolised the vast, dry Australian continent.

McKimson, whose brothers were also animators, worked with Warner Bros. for about 35 years. Jones called him 'one of the greatest',[12] and credited the series of widely used model sheets McKimson drew in the 1940s for the definitive look of the characters in the *Looney Tunes* stable. And 'in his art he was fast, he was fluid, and he was on-the-money'.[13]

There are several explanations for the origin of Taz. The most 'official' is found in the lavish *Warner Bros. Animation Art*:

> Taz auteur Bob McKimson recalled that the character was born when writer Sid Marcus was 'kicking around' different types of characters. And I said, 'About the only thing we haven't used around here is a Tasmanian Devil.' He didn't even know what they were. And we just started talking about it and we came up with this character.[14]

An alternative explanation, found on a number of *Looney Tunes* fan websites, is that McKimson and Marcus created the manic creature as a new test for Bugs Bunny.

Then there is the potential Flynn link:

> *Desperate Journey* is the first of two films in which Errol Flynn actually plays an Australian, which is what he was (Warner PR spread the word that Flynn was Irish in an effort to tone down a wild history. Then again, nobody has yet to either confirm or deny whether he was in fact the inspiration for their Looney Tunes' cartoon creation The Tasmanian Devil [aka Taz]). It is amusing to watch Flynn try to effect a mild Aussie inflection in places, but he eventually gives up and sounds like he usually does.[15]

A 1998 feature article in the *Sunday Tasmanian* produced yet another explanation. The opportunity for the article arose from a visit to Tasmania by Chuck McKimson, Robert's brother, travelling with an exhibition of Warner Bros. artwork:

> Fifty-five years ago in a California art studio two brothers shared morning coffee while solving a crossword. It was a 'regular day' for Warner Bros. animators Robert and Chuck McKimson. Each morning Bob and Chuck would play teasing word puzzles. The crossword ritual primed the talented and successful siblings for a creative day inventing quirky adventures for characters such as Bugs Bunny and Daffy Duck.

One fateful morning back in 1953, Bob was solving a word-clue referring to a spinning creature indigenous to the Australian island state of Tasmania. Bob and Chuck, like most Americans even today, knew little about the heart-shaped island south of the Great Southern Land. But the McKimson boys did know the clue's answer—the Tasmanian devil. The question was common in 1950s crosswords. Bob and Chuck had answered it before. 'Invariably, during that time, the clues would mention Tasmanian devil. They wanted to know what the spinning animal of Tasmania was', Chuck, now 83, recalled. 'My brother, Bob, was a crossword addict and every morning at 9 o'clock he'd sit down to do his crossword puzzle . . . the rest of us did the same thing.'

This particular 1950s morning Bob, Chuck and their fellow Warner Bros. animators were searching for a new character to

Related? Despite the original Warner Bros. cartoon character being created in the 1950s when very little was known about the Tasmanian devil in the United States, these images reveal some intriguing similarities between the real and the invented animal. (Taz courtesy of Warner Bros. Taz, Tasmanian Devil and all related characters and elements are trademarks of and © Warner Bros. Entertainment Inc. Bipedal devil at Bonorong Wildlife Park courtesy The Mercury*)*

> play with their cool-as-a-cucumber rabbit, Bugs Bunny. 'The studio manager wanted a new character and we'd done cats and rats, horses and cows, chickens and whatever . . . so Bob says, "Let's do some research on the Tasmanian devil",'[7] Chuck said. 'So we got encyclopaedias and did some research.'
>
> They learned the Tasmanian devil was a ferocious little creature with a legendary growl and a propensity to run around creating mayhem. 'We looked into how it behaved but there wasn't in fact too much on the Tasmanian devil in those days but whatever there was we went into it', he said. The wild creature would be an exciting counterpoint to and playmate for Bugs, and a team of about five Warner Bros. animators began sketching preliminary drawings. 'All five of us came up with an almost identical looking critter and then my brother, Bob, took those and made the final decision on what it looked like and he made the final drawing', Chuck said.[16]

And so the creature was born.[17] That first six-minute cartoon probably required about 150 story sketches, followed by up to 10 000 images painted on transparent animation cels, the whole process taking some five weeks—call it 20 working days (coincidentally about the same as the real Tasmanian devil's gestation period).

That first experimental cartoon is particularly important, for its own sake and because the 'whirling dervish'[18] very nearly didn't survive. Here is what happens in *Devil May Hare*:

> The animals flee from Taz, who will devour anything and everything, past Bugs's hole, and the wily rabbit tries to bamboozle Taz with a succession of artificial animals he could try to eat. In the end, in desperation, Bugs places a Lonely Hearts ad for a female Tasmanian devil who has matrimony in

> mind. One flies in immediately from, presumably, Tasmania, and Bugs, in the guise of a rabbi (geddit?), marries the pair, thereby calming Taz's savage soul. (The quasi-Freudian equation of Taz's violence with a lack of sex went remarkably unremarked-upon at the time.) As the pair flies off Bugs comments: 'All the world loves a lover, but in this case we'll make an exception.'[19]

Eddie Selzer, an executive producer at Termite Terrace in 1954, objected to the new cartoon character. He thought it too violent for a junior audience, and distasteful to parents. (Warner Bros., through eldest brother Harry in the early decades, had had a strong guiding social principle, believing that cinema could and should morally educate.) Selzer didn't appear to understand the animators. Chuck Jones refers to:

> [the] twelve dreadful years of his reign . . . Perhaps his finest hour came at a story session. Four or five of us were laughing over a storyboard when once again Eddie stood vibrating at the doorway, glaring malevolently at us and our pleasure and laughter. His tiny eyes steely as half-thawed oysters, his wattles trembling like those of a deflated sea cow. 'Just what the hell,' he demanded, 'has all this laughter got to do with the making of animated cartoons?'[20]

Executive producers have power. Selzer ordered that no more devil cartoons be made. Yet someone with greater power saved Taz. Re-enter Jack Warner, who wanted the Tasmanian devil back. It was a curious decision, because that lone cartoon had seemingly been destroyed by time, never mind Selzer. Yet three years later Jack ordered his animation team to make more. And with McKimson directing, that happened—between 1957 and

1964, *Bedevilled Rabbit,* then *Ducking the Devil,* then *Bill of Hare*, then *Dr. Devil and Mr. Hare.*

Dissected by Theodore, admired by Errol, punted on by Jack, the devil almost died but was resurrected again. And the rest, as they say, is history . . .

13

OWNING THE DEVIL: TASMANIA AND WARNER BROS.

In 1960s suburban Melbourne feeding his caged devils was a problem for famed anthropologist and photographer Donald Thomson, until he found a fishmonger at the Victoria Market only too willing to give him leftover fish scraps . . . The arrangement worked well until Donald discovered that his fishmonger friend was a leading member of the Australian Communist Party. This no doubt caused some embarrassment to Donald, considering his good relationship with his other friend, Sir Robert Menzies, and Sir Robert's noted obsession with communist infiltration at that time.

John Teasdale, Rupanyup, Vic.

Taz has no evil machinations. He is not greedy in the sense of wanting power, monetary or political, over others. He just wants to eat, by all and any means. Taz is an innocent savage. He never rose to a civilized state and then reverted. He never fell from grace because he never had it. He has remained in a state of nature as its most powerful force . . . He is so

> outlandish as to not remind viewers of the brutes from which they evolved. Rather, Taz makes the beast of instinct look completely external, lovably innocent, and easy to outwit.[1]

There are tens of thousands of books about the movie industry. Hardly any indulge in academic analysis of animated cartoons. Yet in this short quote it is possible to reflect on Taz the cartoon devil from perspectives of political philosophy, evolution, psychology, and cruelty in humour (we enjoy his stupidity). But of course Taz, along with his maker, has had the last laugh, given the amounts of money he has generated. How did that come about?

Having lain virtually dormant for a quarter of a century, it may seem surprising that such an apparently one-dimensional character was selected by Warner Bros. to become a major production of its new animation studio. In 1990 the company teamed with Steven Spielberg to create syndicated cartoons for television, and the reborn *Taz of Taz-Mania!* appeared the following year with his own 65-episode series. Not only did TV guarantee far greater exposure than film, Taz now had a genuine fictional identity: eighteen years old, a cave home, parents, a brother and sister, a job, a pet, a friend, a rival and an enemy.

A number of other factors lay behind the success. He appealed equally to children, teenagers and adults, and on some networks *Taz of Taz-Mania!* moved into adult time slots. He rode the crest of a new merchandising wave through catalogue shopping and the proliferation of Warner Bros. Studio Stores across the United States and into Europe and Asia. He ranked high in 'pop iconography . . . the rise of animation fandom'.[2]

Real Tasmanians were bound to take notice and 1997 marked the beginning of an interesting relationship between the entertainment giant and the Tasmanian state government. A front-page report in the *Sunday Tasmanian* newspaper, headlined 'We Lose Millions as Yanks Grab Devil', initiated proceedings:

> A multinational company makes millions of dollars out of the Tasmanian Devil—and Tassie does not get a cent. US company Warner Bros. owns the Tasmanian Devil. The international entertainment giant rakes in a fortune from its world-famous cartoon character Taz. The Devil ranks with Bugs Bunny and the Road Runner in the top three most popular characters worldwide—and its sales are increasing dramatically . . .
>
> Warner Bros. International public relations consumer products director Annie Morita said the Tasmanian Devil was one of the rising stars of its merchandising. 'I don't know if I could even attach a number,' she said. 'You'd have to think about everything from home videos, to music to television to studio stores to licensing. He's up there with all of that. You're talking millions if not billions.' She also noted that Taz was even bigger than Bugs in Brazil, Argentina, Mexico and Venezuela because 'they feel he represents the machismo of the region.'[3]

She was not exaggerating. A 1995 Warner Bros. survey had shown that about 95 per cent of US residents recognised Taz. Adult men liked his 'aggressive behaviour', teenagers 'identified with his rebellious streak', and children liked his generally wild manner.[4]

The report also noted a push to link Taz to the Sydney 2000 Olympic Games, as part of a campaign to promote Tasmania internationally. It would require a joint venture between the

Commonwealth government, Tourism Tasmania and Warner Bros., with Taz hosting a virtual tour of the island's tourist draw-cards such as its wilderness areas, wineries and kunanyi/Mount Wellington. For the backers of the concept, the cartoon devil 'attracted the sort of international attention money could not buy. Tasmania should cash in on Warner Bros.' investment'.[5]

Many Tasmanians knew of the cartoon character. Few, however, had any notion that their iconic marsupial could somehow be 'owned'. The report enlightened them:

> Warner Bros. has trademarked the character and registered the name Tasmanian Devil. The patent covers everything from sports gear, dolls, video games and Christmas decorations to underwear. And it is policed. Warner Bros.' Australian legal arm is tracking the source of illicit Tasmanian Devil dolls sold at the Royal Adelaide Show with a view to prosecution. One Tasmanian company, under threat of legal action, battled the international giant for eight years after being told not to use Tasmanian Devil as a brand name. Warner Bros. eventually gave the Tassie firm, Wigston's Lures, a one-off agreement allowing it to call a fishing lure a Tasmanian Devil.
>
> 'We're the only people other than Warner Bros. to have that registration and we fought tooth and nail to get the darn

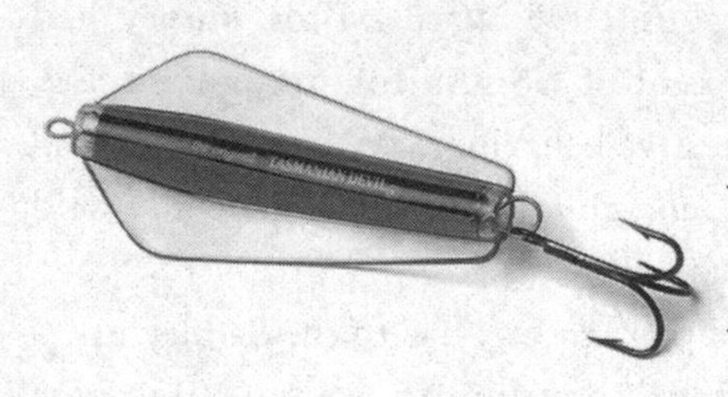

The name 'Tasmanian Devil' is copyrighted to Warner Bros. Wigston's Lures, a small Hobart fishing lure company, spent eight years battling for the right to use the name Tasmanian Devil for one of its lures. Wigston's Tasmanian Devil *is a 13.5 gram 'Beetle Bomb'. (Courtesy Garth Wigston)*

thing,' Stuart Wigston said. 'It's hard to believe something indigenous to Tasmania is registered by a huge, great, multinational company. It's unreal . . .'[6]

The report went further, roping in Tasmania's internationally successful catamaran-building company Incat, which owner Bob Clifford had developed using unique aluminium wave-piercing hulls. It noted that:

Incat Tasmania steered clear of calling its new vessel the Tasmanian Devil—opting for '91-metre Devil'. A disclaimer circulated at the catamaran's launch . . . stated: 'No challenge to Warner Bros.' ownership is intended or implied[:] the devil livery on the 91-metre vessel bears no resemblance to the cartoon character.'[7]

Warner Bros. was asked to respond to the newspaper report. Its Australian lawyer stated that the company pursued offenders:

We've got a duty to go out there and police the marketplace so people don't abuse the system . . . If someone was to use Tasmanian Devil as a trademark we would regard that as an infringement of our registration but it really depends on [the] circumstances in which they used it. If someone was talking about the Tasmanian devil in Tasmania, that's a different situation . . . The trademark is not designed to stop the public from using the expression or to stop someone calling a Tasmanian devil a Tasmanian devil. It's more or less a brand name for goods.[8]

Needless to say the report created political waves. Tasmania's then Tourism Minister, Ray Groom, protested:

It seems so unfair [Groom told the *Sunday Tasmanian*]. Here we are, a small island below Australia with half a million

> people. We've got the devil as a native animal which doesn't exist anywhere else and a big American international company has pinched our rights . . . Warner Bros. are pretty tough operators, they know how to get their copyrights and their trademarks registered around the world where it counts. They endeavour to tie it all up . . . It could finish up in the courts and we'll be looking at that issue as well, to see what we can do to retrieve the rights to use our devil as we want to.[9]

Groom did concede that amicable negotiations were preferred, and to that end a Tasmanian delegation drew up preparations to visit Warner Bros. in Los Angeles for discussions.

Letter-writers to the newspaper didn't hold back. From Marcus Rowell, Hobart:

> I have recently returned from a trip to South America where I was astonished to find that almost everybody knew the Warner Bros. Tasmanian Devil . . . certainly this character Taz is an identity that is internationally known who should be leading Tasmania's international marketing effort. Although we do need to make sure potential visitors know that he is fictitious as some South Americans expressed great concern that we have such a fierce creature in our wilderness.

From Craig Wellington, Hobart:

> It seems to me Warner Bros. took the initiative long before anyone in Tasmania did and made the Tasmanian devil a household name around the world. They have obviously invested vast amounts in their characterisation of the Tasmanian devil and it is understandable that they wish to protect that investment by policing the use of their trademark character. It also seems to me their investment has given the devil a massive international profile. Tasmania should be grateful for

> such a gift. I know it's a cliché in the tourism industry to say 'Look what Crocodile Dundee did for the Northern Territory' but the Tasmanian devil provides us a similar, if not better, opportunity . . . have the Tasmanian devil declared a state emblem (rather than its extinct cousin) . . .

From John Williams, Glebe, Hobart:

> The attitude of Tasmanian politicians that it is all right to rip-off Warner Bros.' trademarked figure, whatever their reasoning, is deplorable. It reinforces the low opinion which the public has of politicians' ethical standards. The reputation of the whole State must also suffer when prospective investors learn that we have such slippery standards in commercial dealings. And anyone who is trying to raise children with a respect for other people's property will be horrified at the example the politicians are setting. Warner Bros. have abided by the system our politicians have set up. I don't know if lawmakers are as careless in other countries. Would, for instance, an Australian company be able to register 'American bald eagle' in the United States of America?

From Barry Giles, Cambridge, Tasmania:

> I recently returned to Tasmania . . . When we arrived in the US about five years ago there was already a vast amount of merchandise in the shops bearing the Tassie devil image . . . most Americans do not realise the animal or the island exist. Actually this is not surprising considering the results of surveys showing their appalling lack of knowledge about the geography of their own continent. You can hardly expect them to know about a distant 'fantasy' island on the other side of the planet. It's a standing joke that Tas . . . Taz . . . is in Africa.[10]

Lively political debate in the Tasmanian parliament followed the government delegation visit to Warner Bros. Groom was pressured to announce a good outcome:

> *Mr Polley:* Can the minister inform the House of the outcome of the negotiations with Warner Bros. and when can we expect to see the internationally-recognised and popular cartoon character being used to promote Tasmania throughout the world?
> *Mr Groom:* Constructive discussions took place and extremely interesting ideas were discussed. I might say contrary to some of the earlier indications the people from Warner Bros. were keen to embrace Tasmania . . . I will not go into the details at the moment because it will be a bit later on when we will discuss this in further detail and make some announcements, one would hope. But they are keen on the environmental aspects of Tasmania—how we can link Taz the Tazzie Devil into promoting internationally environmental issues focusing in part on Tasmania. The indications are this is not going to cost us the millions of dollars we thought . . . I have, Mr Speaker, so many Tazzie Devils coming out of my fax machine—not actual devils but different designs of devils from all sorts of people around Tasmania . . . Some very exciting ideas based upon our wonderful native animal, the Tasmanian devil, others not looking so happy, probably not the sort of thing we would want to use . . . We are looking at these, Mr Speaker, to see how we can use the Tasmanian Devil. It has a lot going for it . . . We are very pleased with the attitude shown by Warner Bros. They appreciate that we have produced this animal—they have used the animal for their own commercial benefits and my argument is that we should gain the benefit, we should be able to use this cartoon

> character based on our own animal widely around the world to promote Tasmania.[11]

Seven months later Groom announced that a verbal deal had been struck, in which the Tasmanian government would pay Warner Bros. an annual fee to use the Taz image for marketing purposes. Speculation on the cost ranged from a low of A$50 000 to a high of ten times that amount, with all sorts of possibilities enthusiastically rumoured: Taz would feature in the opening ceremony of the Sydney 2000 Olympics; he would appear on Tasmanian tourist brochures in the United States, Canada, Europe and Asia; and a human in a Taz suit would become a feature of major world events such as the Berlin International Travel Show and the Melbourne Grand Prix.

In the 1990s the Tasmanian government lobbied Warner Bros. to allow use of the Taz image to promote Tasmania internationally. Warner Bros. declined. (Taz courtesy of Warner Bros. Taz, Tasmanian Devil and all related characters and elements are trademarks of and © Warner Bros. Entertainment Inc.)

Groom enthused particularly about the Games, declaring Taz to be integral to a Tasmanian push so significant that 'it's developing bigger than Ben Hur', though he did concede that 'whether we can get him on the main arena in the opening ceremony is a tough one'—suggesting a battle with the Games organisers—and furthermore, that the battle with Warner Bros. wasn't quite over either: 'No contracts are signed yet but we have understandings'.[12]

Whatever those understandings were, they dissolved, and the Games duly went ahead without Taz. But the issue was about more than the mere commercial opportunism represented by the cartoon figure, because its real counterpart represented an idealised opposite: a rare and elusive creature inhabiting an unspoilt wilderness.

During the parliamentary debates the Tasmanian Greens' Peg Putt had pointed this out. In 1997 debate raged over the conservative Federal Government's proposed Regional Forest Agreement (RFA)—would it protect or destroy Tasmania's old growth forests?—and Putt stated the obvious:

> Tourism is booming. Tourism is providing more and more jobs and tourism to Tasmania is promoted very much on the basis of 'Tasmania our natural State'. Our distinguishing feature in the international market is our wilderness and our beautiful places, our forests and our World Heritage areas, and if the RFA fails to protect that resource for the tourism industry then it will have failed. Only today we have had the announcement about Warner Bros. looking at promoting Tasmania with the Tasmanian devil focusing particularly on environmental protection. That is where a big-jobs future for our native forests lies.[13]

Warner Bros. appeared to have no particular history of championing distant environmental causes. Why should it stipulate that its cartoon character be linked to 'environmental issues focusing in part on Tasmania' (to quote Groom)? The Tasmanian Government–Warner Bros. verbal agreement, whatever it had been or not been, evaporated.

14

DEVIL FACIAL TUMOUR DISEASE

Scenario: without warning, a vicious disease unknown to modern science targets one of the world's rarest animals—so vicious it feasts on the face and mouth before invading the body, taking control of the vital organs. Death is within months through starvation. This animal, long holding out on a remote island, already has a small population and low genetic diversity. The disease methodically wipes out more than 80 per cent of them. How and why did this happen? And, most importantly of all: what chance is there of survival against such huge odds? None, you would think. But no, here comes the legendary, indestructible devil.

This is not fiction tarted up as doomsday bioscience, but reality. It is devil facial tumour disease (DFTD). Is it a harbinger, a precursor, of what lies ahead in our world? This is a disturbing possibility, not least because although the disease broke out about 30 years ago, it remains incurable. This is despite intensive worldwide pathology and diagnostic studies, and the development of various critical immune strategies. Equally disturbing is that the disease type is incredibly rare; it is caused by a transmissible cancer, not a virus. Nor is its originating trigger known.

Noted US author and journalist David Quammen introduces his 2012 book, *Spillover: Animal Infections and the Next Human Pandemic*, with these words: 'As globalization spreads and as we destroy the ancient ecosystems, we encounter strange and dangerous infections that originate in animals but that can be transmitted to humans. Diseases that were contained are being set free and the results are potentially catastrophic.'[1]

Not only was this book prophetic vis-à-vis the COVID-19 pandemic, which occurred seven years after its publication, this fundamental premise may also apply to DFTD. Was the Tasmanian devil disease sparked by some anthropogenic trigger in the animal's habitat, or even a species spillover? DFTD is spread by one Tasmanian devil transferring infected tumour cells to another, through agonistic biting and claws-scratching. Biting is fundamental to devil behaviour—while it is the cause of the spread, biting is not the reason for the disease.

Well into the ultra-smart 21st century, science on the run has been a feature of this mysterious disease since its discovery in the 1990s. Very few other transmissible cancers are known. One occurs in dogs (canine venereal cancer) and others occur in some species of marine bivalves—clams, mussels and cockles. The origin of DFTD is in the devil itself; it has arisen twice in the form of two different cancer lineages, called DFT1 and DFT2, in the 'founder devils', which then spread the diseases to other devils. The tumours in all affected devils are clones with an identical genotype to the original devil in which they arose—in the case of DFT1, a female devil in northeast Tasmania and in the case of DFT2, a male devil in the southeast. Some devils have both forms of DFTD.

The actual tumour derives from what is known as a Schwann cell, which is a cell enabling peripheral nerves to function properly. When the body has a wound, Schwann cells at the site multiply and repair the wound, then stop multiplying as the wound heals. For some reason in the founder devil, instead of repairing its facial wound, the Schwann cells generated a cancerous tumour. What might have caused this in the founder devil? One plausible theory is that the 'combined effect of factors such as frequent nerve damage from biting, Schwann cell plasticity and low genetic diversity may allow these cancers to develop on rare occasions'.[2]

As the disease spread among the devils, the population crashed. By 2020, when the disease had reached about 90 per cent of the devils' range, the wild population had fallen from an estimated 53,000 in 1996 to less than 17,000. Some local declines peaked at more than 80 per cent, enough to trigger biological extinction, or an inability for the species to recover. In the first decade or so, species extinction was genuinely feared. But genomic research shows that the rate of increase of infections has slowed naturally. Together with high-level human intervention, this means the Tasmanian devil extinction scenario is unlikely—at least for now.

Imagine if DFTD with the same mortality rate had occurred say, 100 years ago, when there was limited or no human intervention, and the animal was viewed as a menace—what chance of survival for the devils then?

Huge efforts have gone into avoiding the devil's extinction. The overarching body is the Save the Tasmanian Devil Program, set up in 2003 as a joint Tasmanian and Australian Federal Government initiative. Later, Australia listed the devil as

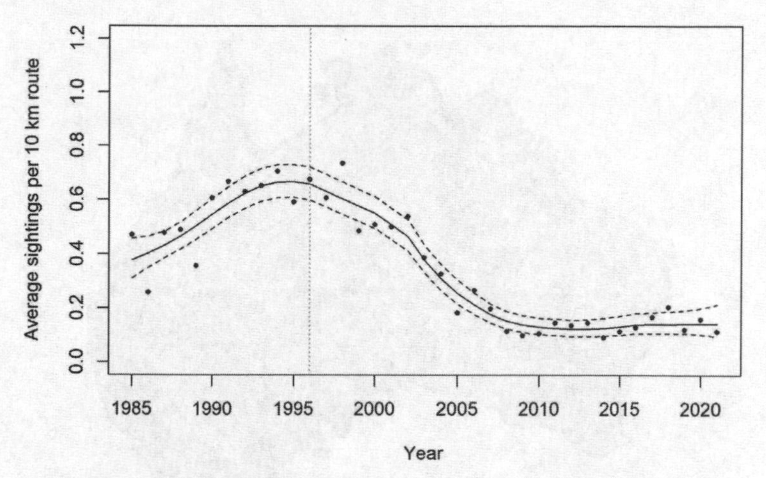

Despite genuine fears that DFTD would cause extinction, it has not. This graph of yearly spotlight surveys, conducted by Tasmania's Department of Natural Resources and Environment, clearly shows DFTD's devastation of the devil population but survival in spite of that collapse. The graph depicts a generalised additive model of the spotlight counts of devils through time showing: predicted trend (solid line), upper and lower 95% confidence intervals (dashed lines), actual observation points (black dots) and the observed year of emergence of DFTD (vertical grey line). (Courtesy Billie Lazenby, Department of Natural Resources and Environment Tasmania)

Vulnerable and then Threatened—it is also listed as Endangered on the International Union for the Conservation of Nature (IUCN) Red List.

The shocking speed at which the disease spread across Tasmania meant that the Save the Tasmanian Devil Program had to put multiple critical priorities in place: caring for and euthanising diseased individuals, many with live pouch young; managing the disease in the wild; and establishing an insurance population. They arranged partnerships with Australia's Zoo and Aquarium Association, and with quality research institutions including the Menzies Institute for Medical Research in Hobart,

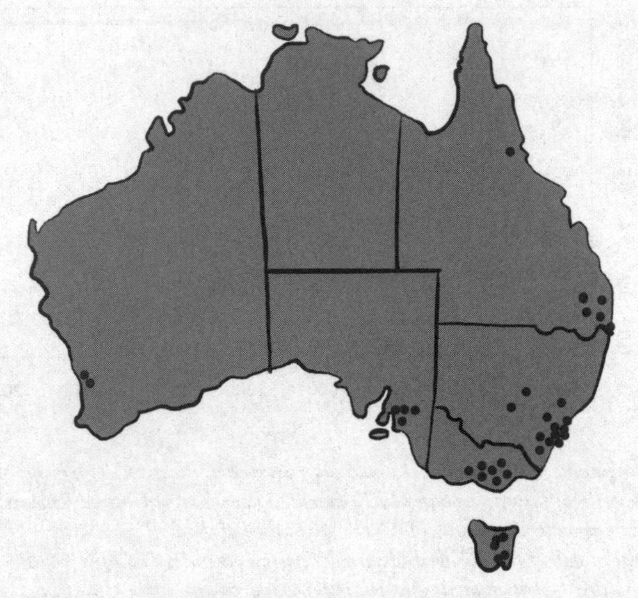

The Zoo and Aquarium Association Australasia (ZAA) represents zoos and aquariums across Australia, New Zealand, Papua New Guinea and South Pacific Islands. ZAA has supported the Save the Tasmanian Devil Program and, in particular, the captive breeding population, known as the insurance population, in order to safeguard the species against extinction. The map indicates the locations of the many ZAA welfare accredited zoos and wildlife parks that contribute to this conservation initiative, which began in 2006.

University of Tasmania, The University of Sydney, University of New South Wales, Macquarie University and University of Cambridge in the UK. Zoos Victoria also became an active collaborator.

A unique international program got underway in 2014 when the Tasmanian Devil Ambassador Program was established: 'International zoos, though an expression-of-interest process, were given the opportunity to exhibit Tasmanian devils as advocacy animals to increase global awareness of the plight of the devils. In return, the receiving zoo was invited to play an active part in *in situ* conservation of the devil in Tasmania.'[3]

The likely saviour of the devil from extinction would be a disease-free insurance population. But how to achieve such an outcome with the disease running rampant? Two Tasmanian localities were chosen (from many considered): Maria Island, adjacent to the mid-east coast, and the Forestier Peninsula southeast of Hobart. Healthy devils were introduced into these localities after being selected from zoo populations and parts of Tasmania in the north, northeast, west and far south, then known to be disease-free.

In quarantined localities, the initial insurance population, known as the 'metapopulation', had commenced a breeding program in 2006. The first release onto Maria Island, of 14 devils, took place in November 2012. In 2013, by the end of the first mating season, all adult female founders had pouch young, with 24 young devils born and surviving to independence. Second and third release programs took place in 2013 and 2017, respectively, to boost the genetic diversity of the first release group.

The carefully managed Maria Island program worked well. Released animals all survived and females bred young in an area with a wide availability of resources—prey and scavenged—and with no competition. The devils also spread out across almost all of the island. Furthermore, the released individuals had been carefully selected for their behavioural traits—for instance, minimal propensity to show unwarranted aggression towards people—and those traits persisted through the offspring.

Maria Island devils have become a healthy, DFTD-free source population of reasonable genetic diversity for reintroduction to mainland Tasmania. A particular success was the release in 2016 of 33 devils, including 32 pouch young, into the wukalina/Mount William National Park in northeast Tasmania. Their survival rate was 100 per cent. Somewhat famously, a devil called Nutella, born and raised in South Australia's Monarto Zoo, was selected for release onto Maria Island. Here she bred, then was selected again for release the following year at another northern Tasmanian location, where she continued breeding until after the age of five.

The Nutella story is a marvellous instance of human–animal interaction, an unlikely good-news story emanating out of an emergency. As was reported under the headline 'Tasmanian Devils set free in military training area': 'Navy, Army, Air Force and Defence Public Service personnel joined forces in a unique effort to support the release of 33 captive Tasmanian Devils into the wild at the Stony Head Military Training Area.'[4]

The release was coordinated by Department of Defence Regional Environment and Sustainability Officer Kate Hibbert, and Warrant Officer Two Robert Butchart, along with Samantha Fox of the Save the Tasmanian Devil Program. Members of the Australian Defence Force helped to set up the release sites across the 5500-hectare property, which provides terrain for military training, including flat rural land, undulating coastal scrub and hilly open forest.

Dr Hibbert said:

> Although at first it seems like the protection of the Tasmanian Devil population is not vital to supporting capability, having a healthy training area is. Without the devils there is increased opportunity for feral animals like cats and rabbits to thrive. This then affects other animals and plants and can result in

Nutella being released into the Stony Head Military Training Area.
*(*Royal Australian Navy News*, 22 September, 2016)*

> an environment that is no longer suitable for training. Stony Head was chosen as the release site because of its remote location and healthy natural environment, which is a great compliment to Defence's land management.[5]

An earlier chapter of this book mentioned a far older military connection. During World War I, Bluey Thompson, the Tasmanian soldier-cook, kept a Tasmanian devil ('Bluey') as a mascot while he served in Egypt with his 12th/40th Royal Tasmanian Regiment. (Bluey was eventually given to Cairo Zoo.) That regiment slowly went into abeyance and was not reconstituted until 1987, as part of which, a key mascot symbol of its earlier existence was enlisted: Bluey Devil. (Note the Army Number!)

As successful as it is, the devil release program has not been without problems. On Maria Island, the healthy new devil cohort exterminated an entire population of some 3000 little

PERSONAL PARTICULARS

Page 1

Surname DEVIL Other Names BLUEY

Army Number TX 666 Corps RAINF Identity Card Number (in full)

Date of Birth 05 NOV 89 Place of Birth BONORONG BRIGGS RD BRIGHTON TAS 7030 (Town/State)

Religion If Will lodged and where

Complexion DARK Marks or Scars WHITE MARKINGS CHEST/RUMP DIAGONAL SCAR ON NOSE

Hair BLACK/WHITE

Eyes DARK BROWN

Height 10"

Weight (lbs)

Physical Disabilities and Special Characteristics (eg, impediment of speech, allergies)

Date of Photograph August 1990

Marital Status SINGLE Blood Group

Signature

Bluey Devil's 'enlistment' details as the official Regiment mascot. Note the Army Number! The Royal Tasmanian Regiment Collection. (Courtesy Colonel Denis Townsend (Retd), Malcolm McWilliams and Andrea Gerrard)

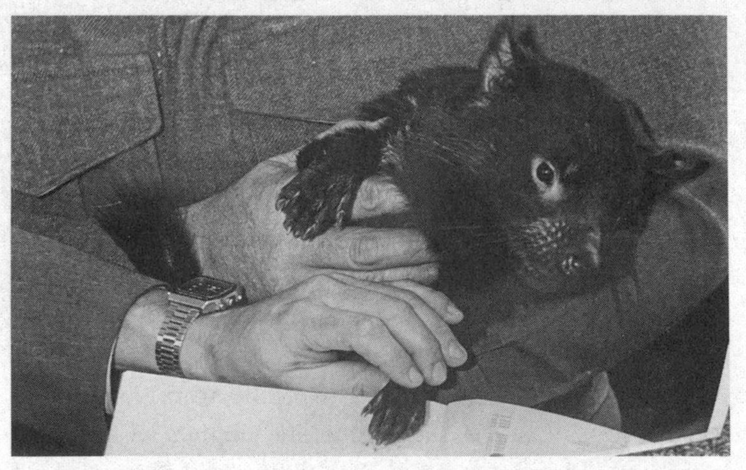

Bluey Devil needed some help inking his paw print 'signature' onto his official enlistment papers. (The Royal Tasmanian Regiment Collection)

penguins; they continue to prey considerably upon wombats and Cape Barren geese on the island, both of which are also insurance populations. Short-tailed shearwaters and native hens also feature in their diet. While this trade-off had been anticipated, it was perhaps not to such an extent.

The Forestier Peninsula project occurred because that large area is connected to mainland Tasmania by an 800-metre-wide isthmus; it is effectively an 'island'. Unlike the Maria program (where there had been no devils previously), during May and June 2012 trapping teams removed 35 Forestier Peninsula devils, both healthy and diseased individuals. This depopulation effort worked so well that the area became devil-free and thus disease-free. In late 2015 and early 2016, 49 devils were released onto the Peninsula (after devil-deterring fence and lighting were erected at the isthmus). But this also came at a cost: 16 of those devils were killed on the roads. The project confirmed that devils bred in captivity and released into the wild are more vulnerable to becoming roadkill than naturally more cryptic/wary wild-born devils.

Together with these carefully managed release programs, the research component of the disease continues to be of critical importance. After many trial-and-error studies, candidate vaccines and immunotherapeutic (immunity-boosting) approaches have been developed, which has resulted in encouraging tumour regression in some DFTD devils. Almost all the immunised devils released at Stony Head and Narawntapu National Park had 'produced specific anti-DFTD antibodies . . . The responses were durable, lasting at least several months after animal release . . . These recent findings . . . and the demonstration of naturally occurring immune responses to DFTD in a proportion of wild devils, provide a firm basis for pursuing an immune solution to DFTD. They indicate that the majority of devils can mount an immune response against DFTD cells, paving the way for an immune solution to this devastating transmissible cancer.'[6]

Zoos continue to be active and enthusiastic participants in devil recovery management, which are effective and important. In Australia, 30 display zoos and wildlife parks initially signed up to participate in growing an insurance population. Internationally, the Tasmanian Devil Ambassador Program places devils in overseas zoos to highlight the precarious status of the species. The Ambassador Program rationale is this:

> Globalisation was determined to be the most strategically, financially and ethically sound response to the high performance of the Australian zoo-based insurance population. It would allow the insurance population within Australia to meet its genetic goals, without the need to seek further government funding to support the expanding housing costs [in Australia]. Globalisation would provide opportunities for the international zoo community to appropriately support the insurance population in Australia and the *in situ* conservation actions of the Save the Tasmanian Devil Program.[7]

In three examples of such support, devil-related income from Prague Zoo helps to pay for the Maria Island work, San Diego Zoo for genetic research, and Toledo Zoo for in-the-field monitoring. The Ambassador Program partners with 19 international zoos in 10 countries. Given the disease's unique nature and our great need to know as much about it as possible—should it happen anywhere else, at any time, to any vulnerable species—this international collaboration is essential. Ultimately, the Tasmanian Government is responsible because it is the legal owner of every single Tasmanian devil in the world.

Even this global effort is not without some controversy. Zoos in Europe, the USA and elsewhere do not breed their Tasmanian devils, because the Australian Government does not allow them to breed devils. They are displayed for awareness purposes only, sent overseas as surplus stock from Australia's breeding institutions; otherwise—according to a poster on the European online forum *Zoo Chat* in 2020—'if Europeans start breeding this will clog up the Australian breeding institutions with unwanted animals'.[8]

Another poster confirmed this belief:

> The entire point of sending older surplus devils out of Australia is to free up space in Australia's virus-free breeding programme for *saving the species from extinction*. Having overseas zoos filling available spaces with home-bred devils *detracts* from the conservation effort in Australia because then there would be nowhere for the older devils to be placed and they would then be using valuable spaces which should be being used by breeders.[9]

Some confusion arose because one European zoo—Copenhagen's—does have a devil-breeding program. But this program predated the Ambassador Program and has nothing to do with the disease. It began soon after devils were gifted to the zoo to celebrate the 2004 marriage of Tasmanian-born and educated Crown Princess Mary to Crown Prince Frederick.

One question put to the *Zoo Chat* forum seemed logical: 'Is there a good reason for not running it as an international program with breeders in Australia and abroad alike?' Another poster agreed: 'Personally, I think the Tassie devil needs all the help it can get in allowing its populations to recover ex situ (and not just in-country). This is true for both non-breeding animals and those designated allowed to breed.' These posts were met with this response, however: 'Afraid I have to disagree with you. There is no conservation value at all in Prague Zoo [or others] breeding devils. If they breed devils it would be entirely for their own purposes. If they want to do something for conservation they should work on one of the more than 10,000 other threatened species on the IUCN list that are not getting the attention that Tasmanian devils are.' This one provided more context: 'If you think it is difficult getting animals out of Australia you should try to get them in. I can't imagine devils from Europe ever being imported into Australia . . . Frankly breeding them is simply an ego trip for the management of the zoo concerned. Running an international breeding program would be a waste of time and money for everybody concerned.'

Greg Irons, the director of Bonorong Wildlife Sanctuary, near Hobart, is greatly experienced in captive devil management. It's not just the disease hammering devils. As he described in 2023:

> Sadly we see them through our rescue service far more often than people realise. In fact as I write this, we have just had the most calls we have ever had over a three-month period with 80 calls that are devil-related, scattered among thousands of others—each individual call [is] an animal suffering, distressed and often in pain and sadly 97 per cent of the calls we receive are deaths or injuries caused by people, either directly or indirectly. Not just a number, but an individual animal in often heart-wrenching circumstances. With our

> devils it isn't just disease. It is dog attacks, vehicles, secondary poisoning, rubbish, barbed wire fencing, stuck in built-up areas, machinery injuries, just to name a few. While most of us cannot assist in fighting the disease, what we can all do is be aware of what causes any animal injury. We must analyse and educate: what can we do differently to avoid them needing assistance? Every life counts—and who knows which one that loses its life wasn't a key individual devil that may have developed some sort of immunity that could help save an entire species? In rehabilitation, Tasmanian devils are one of the toughest animals I have come across, but the numbers can't keep going the way they are.[10]

Greg Irons, like Jack Bauer in an earlier chapter, intimately understands the devil's characteristics. He writes of the honour he feels

> at rescuing wildlife and helping them return to the wild, while providing a sanctuary and a safe home for those that cannot be released due to injuries and other circumstances after meticulous assessment to make sure each individual is suited to a life with us and can live a happy and meaningful existence. The Tasmanian Devil is an animal that taught me that no matter how much we think we know and understand there is always more to learn. It was only after many years that I truly got to know these hugely misunderstood animals. There are cheeky devils, sneaky ones, shy ones, over-confident ones, hyperactive ones. Chomper, PeeWee, Dougie and Griffen were the first four Tasmanian devils I raised, when they were just starting to get a 'stubble look' as their fur started to pierce through their pink skin. Their eyes had only just opened and they were barely capable of moving. As they grew, quickly, and recognised my smell and sound, they'd excitedly vocalise

> and climb onto me wherever they could, gently biting on to my clothes and latching before promptly falling asleep in a state of protected calmness and achievement. They did this regularly through to adulthood—loving, affectionate, gentle, sweet and vulnerable natures, that most people don't associate with this creature, promoted as it is with a mouth wide open bearing its teeth, or eating a carcass after a violent clash with one of its own kind. Chomper: needy, cuddly little lady. PeeWee: runt of the four, cheeky, affectionate, shy. Dougie: goofy, over-confident. Griffen: quiet leader watching everything unfold. So when we look at these strong, seemingly indestructible creatures and their fearsome screams and jaw strength—take a moment to remember, just like us, they're the full suite of distinct personality traits.[11]

This chapter concludes a discussion of 'rewilding'. This is defined as 'a progressive approach to conservation. It's about letting nature take care of itself, enabling natural processes to shape land and sea, repair damaged ecosystems and restore degraded landscapes. Through rewilding, wildlife's natural rhythms create wilder, more biodiverse habitats.'[12] The Tasmanian devil represents each polar opposite of this definition, because for the species to be 'rewilded' back into a natural state, it has had to be humanly micromanaged to within an inch of its life. Other species have been successfully rewilded, such as wolves in Yellowstone National Park, beavers in Scotland and bison in Europe, but these projects did not have existential species doom as the prime motivator. In Tasmania, wildlife biologists on the ground have been rewilding devils in their natural habitat for many years through the Wild Devil Recovery Project, which is 'aimed at supporting depleted wild populations across Tasmania through the translocation of healthy devils into diminished populations'.[13]

The concept of rewilding the Tasmanian devils extends to reintroducing them to mainland Australia, their former habitat. Might this be a logical step to take, not only deferring species extinction but being truly rewilding? It's a tantalising idea but flimsy. The devil went extinct on the mainland 3000 years ago. How far back do we go to justify reintroductions—to Neanderthals, woolly mammoths? The reality is, we are still arguing about why the devil (and thylacine) went extinct on the mainland, so we don't fully understand what will happen if it is reintroduced. We don't know what impacts a scavenging hunter with a taste for all animal proteins will have. We could potentially introduce DFTD to the mainland. It is the Tasmanian devil, too, not the Mainland devil.

Is there some additional means of protecting the Tasmanian devil into the future? David Quammen and many others know the problems caused for wild animals through habitat fragmentation; there is also a possible anthropogenic trigger in the environment contributing to the devil's disease. Quammen showed, for example, that habitat fragmentation was the Hendra virus disease trigger. In Queensland, bats hosting the Hendra virus were dispersed due to their original roosting sites being fragmented, and they ended up roosting in great numbers above areas where racehorses lived. The virus passed into the horses and from them to their trainers, causing multiple human and horse deaths.

Devils, by virtue of their relatively small home ranges and broad foraging techniques, live well in fragmented habitat provided they have suitable den sites such as overhangs and large

TASMANIAN DEVIL

In nature, as well as in name, the Tasmanian Devil, *Sarcophilus harrisii*, is an unpleasant brute. It is a marsupial, carrying its young in a pouch as the kangaroo and the possum do, but instead of feeding like kangaroos and possums on vegetation, it is exclusively a meat eater. Moreover, it is not "choosy" about its food—stale meat that would disgust the nose of a pariah dog is still good food to the devil, though a fresh kill in a fowl-yard is his delight under modern conditions of civilization. It is for that reason, particularly, that the hand of every farmer in Tasmania is against him, and the Devil is fighting a losing battle and falling back into the fastnesses of the wild, unexplored, "horizontal" country.

But in spite of all the unpleasantness of his nature, and the economic loss which he occasionally causes, there are reasons why we should protect even the Devil. He is biologically and anatomically unique. He is the sole surviving member of an interesting animal family which used to roam much of the world in ancient times, and Tasmania is his last stronghold. Even though his protection be confined to the limits of the large national park reserves with which Tasmania is blest, let him be protected as a living museum piece, for if he goes from there he goes from the world for ever.

Advertisement of Carlton & United Breweries Ltd., Makers of Victoria Bitter, a healthful beverage.

CV.19.10

This compelling illustration of a devil, which appeared in a 1950 advertisement for Carlton and United Breweries as no. 19 in their 'Save Our Native Fauna' series, is notable for its relatively early conservation message from a commercial enterprise. (Weekly Times, *Melbourne, Wednesday 14 June 1950, page 48.)*

fallen trees; but there is another catch. When their numbers swell, the high densities of devils increase the likelihood of multiple disease outbreaks. There is also the reality of roadkill. Between January 2021 and March 2022, Tasmanian wildlife campaigner Alice Carson counted a total of 114 dead devils along a 25-kilometre stretch of road in Woolnorth, the 6000-hectare farm and tourist destination in the state's far northwest.[14]

Potential further protection could therefore come in creating a dedicated Carnivore Park in Tasmania, a large area with good amounts of suitable habitat and minimal human presence. In 1990, mammalian expert John Eisenberg of the University of Florida visited Tasmania, and proposed such a park. Eisenberg, 'whose research and writing on animal behavior, genetics and evolution made him one of the world's foremost experts on mammals' according to *The New York Times*,[15] discussed with David Pemberton and others the need for species conservation via a unique marsupial guild in Tasmania.

To that end, he prepared an unpublished manuscript, titled 'Status and Conservation Prospects for the Larger Carnivorous Marsupials of Tasmania', for the Florida Museum of Natural History. He identified that, while 50 per cent of Tasmanian land is reserved, most is not good habitat for larger carnivorous marsupials; of what reserved lands there are, they're too small to sustain genetically independent populations. His metrics were a population of 500 adults; for large carnivores, he calculated they would require a Carnivore Park of 10,000 km^2 for the thylacine, 833 km^2 for the Tasmanian devil, 2300 km^2 for the spotted-tailed quoll and 1000 km^2 for the eastern quoll. His dream was one contiguous area of suitable habitat for the carnivore guild. The large tracts of wilderness in the northwest, known as takayna/Tarkine, might prove suitable. (Ironically, had the thylacine survived, it would have probably been in that area.)

This book's Introduction mentions the late Geoff King's devil restaurant at Marrawah, at the northern coastal tip of takayna/Tarkine. This could be a fitting way for respectful human visitors to enter a permanent safe home for the long-suffering Tasmanian devil.

NOTES

Introduction

1 Loh, Dr Richmond, 'Tasmanian Devil (*Sarcophilus harrisii*) Facial Tumour (DFT)', paper prepared for the 14 October 2003 Launceston workshop, p. 2.

Chapter 1

1 Letter of Morton Allport to Curzon Allport, March 1863. Allport Library & Museum of Fine Arts.
2 Definitive studies were carried out by Lars Werdelin and documented in his 'Some Observations on *Sarcophilus laniarius* and the evolution of *Sarcophilus*', Records of the Queen Victoria Museum, vol. 90, 1987.
3 Nick Mooney, occasional taxonomist.
4 Scott, Alan, pers. comm., 1 July 2004.
5 ibid.
6 Guiler, Eric, *The Tasmanian Devil*, Hobart, St David's Park Publishing, 1992, p. 10.
7 *The Mercury*, 2 September 2003, p. 2.
8 Nowak, Ronald M., *Walker's Mammals of the World*, Baltimore, Johns Hopkins University Press, 1999, p. 64.
9 Nick Mooney email to David Pemberton, 8 November 2004.
10 Wilkie, A. A. W., as told to Osborn, A. R., 'Tasmanian Devils[:] Three Interesting Imps', in *Reminiscences From the Melbourne Zoo*, Melbourne, Whitcombe & Tombs, 1917, pp. 58–9.

11 Lord, Clive, 'Notes on the Mammals of Tasmania', in *Royal Society of Tasmania{:} Papers and Proceedings, 1918*, Hobart, The Society, 1918, p. 45.

12 Lord, Clive, 'Existing Tasmanian Marsupials', in *Royal Society of Tasmania{:} Papers and Proceedings, 1927*, Hobart, The Society, 1927, p. 22.

13 www.jonahcohen.com/jersey_devil. This website and many others are devoted to information about the State of New Jersey and its famous devil.

14 Lord, Clive, *A Synopsis of the Vertebrate Animals of Tasmania*, London, Oldham, Beddome & Meredith, 1924, p. [ii].

15 Cameron, Max, pers. comm., June 2004.

16 Fleay, David, 'The Tasmanian or Marsupial Devil—Its Habits and Family Life', *The Australian Museum Magazine*, vol. X, no. 9, 15 March 1952, p. 277–8.

17 Linnean Society of London, *Transactions*, vol. 9, 1808. 'Description of two new Species of Didelphis from Van Diemen's Land. By G. P. Harris, Esq. Communicated by the Right Honourable Sir Joseph Banks, Bart. K. B. Pres. R. S., H. M. L. S. Read April 21, 1807', reproduced in *Letters of GP Harris 1803–1812*, edited by Barbara Hamilton-Arnold, London, Arden Press, 1994, p. 90.

18 Fleay, David, op. cit., p. 279.

19 Wilkie, A. A. W., op. cit., pp. 58–9.

20 Grzimek, Bernhard, *Australians{:} Adventures with Animals and Men in Australia*, translated by J. Maxwell Brownjohn, London and Sydney, Collins, 1967, p. 278.

21 Fleay, op. cit., p. 278.

22 Guiler, op. cit., p. 18.

23 www.dpiwe.tas.gov.au/inter.nsf, p. 7, accessed 30 December 2003.

24 Grey, Lionel, pers. comm., 10 July 2004.

25 ABC Radio, *PM*, 2 December 2002. www.abc.net.au/pm, accessed 8 March 2004.

26 www.web.macam98.ac.il, accessed 10 March 2004.

27 Fleay, op. cit., p. 277.

Chapter 2

1 Long, John, Archer, Michael, Flannery, Timothy, and Hand, Suzanne, *Prehistoric Mammals of Australia and New Guinea: One Hundred Million Years of Evolution*, Baltimore, Johns Hopkins University Press, 2002, p. 32.

2 ibid., p. 55.

3 Australian Museum Online, accessed 4 January 2004. www.amonline.net.au/webinabox/fossils

4 Michael Archer, Suzanne Hand, Suzanne J. Hand, Henk Godthelp, *Australia's Lost World: Prehistoric Animals of Riversleigh*, Indiana University Press, 2000, Foreword.

5 Long et al., op. cit., p. 55.

6 www.environment.sa.gov.au/parks/naracoorte, accessed 5 January 2004.

7 Wroe, Stephen, 'The Myth of Reptilian Domination', *Nature Australia*, Summer 2003–2004, p. 59.

8 Morrison, Reg, and Morrison, Maggie, *The Voyage of the Great Southern Ark*, Sydney, Lansdowne Press, 1988, p. 292.

9 Tasmanian evidence is instructive here. La Trobe University academic Dr Richard Cosgrove, a specialist in late Pleistocene archaeology, examined over 48 000 bones from middens and cultural sites across southwest Tasmania. They were overwhelmingly made up of Bennett's wallaby and wombat, the major Aboriginal food items for over 20 000 years. Just fourteen devil bones were found. That rules out any notion of overkill and instead emphasises good harvesting management. Cosgrove's work also found no evidence of human predation on megafauna, suggesting that they were extinct before human arrival at the southeast tip of the Australian continent and therefore succumbed to something other than overkill.

10 Based on analysis of a limestone hammer by Charles Dortch, Curator of Archaeology at the Western Australian Museum, using enhanced radiocarbon dating and optically stimulated luminescence methods.
11 Gill, Edmund D., 'The Australian Aborigines and the Tasmanian Devil', *Mankind*, 8 (1971), p. 59.
12 Noetling, Fritz, 'The Food of the Tasmanian Aborigines', *Papers & Proceedings of the Royal Society of Tasmania,* 1910, p. 281.
13 Flood, Josephine, *Archaeology of the Dreamtime*, Sydney, Collins, 1983, p. 62.

Chapter 3

1 Jones, Menna, 'Convergence in Ecomorphology and Guild Structure among Marsupial and Placental Carnivores', in Jones, Menna, Dickman, Chris and Archer, Mike (eds), *Predators with Pouches: The Biology of Carnivorous Marsupials*, Collingwood, Vic, CSIRO, 2003, p. 290. She cautions, however, that the success rate of such attacks is unknown.
2 ibid.
3 ibid.
4 Ewer, R. F., *The Carnivores*, London, Weidenfeld & Nicolson, 1973, p. 76.
5 Lord, Clive, 'Existing Tasmanian Marsupials', op. cit., 1927, p. 22.
6 www.wolverinefoundation.org, accessed 30 January 2005.
7 ibid.
8 ibid.
9 ibid.
10 www.napak.com/honey_badger, accessed 31 January 2005.
11 www.awf.org/wildlives/183, accessed 30 January 2005.
12 ibid.
13 ibid.
14 ibid.

15 https://harari.gov.et/ao/landmarks/hyena-feeding-
16 Eisenberg, J.F., *The Mammalian Radiations*, Chicago, Ill., University of Chicago Press, 1981.
17 Menna Jones interview with David Owen, 1 October 2004.
18 ibid.
19 Strahan, Ronald (ed.), *The Mammals of Australia*, rev. edn, Chatswood, Reed Books, 1995, p. 60.

Chapter 4

1 www.abc.net.au/science/scribblygum, accessed 30 December 2003.
2 Fleay, David, 'The Tasmanian or Marsupial Devil—Its Habits and Family Life', op. cit., pp. 279–80.
3 Gilbert, Bill, *In God's Countries*, Omaha, University of Nebraska Press, 1984, p. 8. Gilbert earned considerable respect as a popular conservation and natural history writer and he travelled to Tasmania specifically to write the eighteen-page chapter on devils which appears in this book. He spoke to a number of people who could readily claim to know much about the devil.
4 Pemberton, David, 'Social Organisation and Behaviour of the Tasmanian Devil, *Sarcophilus harrisii*', thesis submitted in fulfilment of the requirements for the degree of Doctor of Philosophy in the Science Faculty, Zoology Department, Hobart, University of Tasmania, July 1990, p. 123. A total of 3788 traps were set in ten sessions for individual devil identification, trapping 328 males and 353 females, 554 and 515 times respectively. Most devils became trap-shy but a few were caught many times. In respect of feeding, wallaby and wombat carcasses 'were placed in a paddock approximately fifteen metres from the edge of the tea-tree scrub running along a creek in the south of the study area. A hide was positioned fifteen metres from the carcass. The carcasses were always c. twenty kilograms in weight and were tied with thin wire to a stake embedded in the ground to prevent

animals dragging them away. Lights were set up on the left and right hand side of the carcass to reduce the amount of light shining directly at the observer or the animals which usually approached the carcass from the bush edge . . . No animals left the carcass site when lights were switched on, and soon after intense interactions began there were animals moving within the white light, around the hide, and through the hide under the observer's chair' (p. 111).

5 ibid., p. 117. The 'yip' was identified subsequent to the completion of the thesis. Thylacines also had a 'yip' call.

6 ibid., p. 164.

Chapter 5

1 Harris, George Prideaux, 'Description of two new Species of Didelphis from Van Diemen's Land. By G. P. Harris, Esq. Communicated by the Right Honourable Sir Joseph Banks, Bart. K. B. Pres. R. S., H. M. L. S. Read April 21, 1807', in Linnean Society of London, *Transactions*, vol 9, 1808. X1, reproduced in *Letters of GP Harris 1803–1812*, edited by Barbara Hamilton-Arnold, London, Arden Press, 1994, p. 90.

2 Gould, John, *Mammals of Australia*, 1863, quoted in Joan M. Dixon (ed.), *The Best of Gould's Mammals*, Sydney, Macmillan, (rev. edn) 1984, p. 44.

3 Meredith, Louisa Anne, *Tasmanian Friends and Foes: Feathered, Furred and Finned; A Family Chronicle of Country Life, Natural History, and Veritable Adventure*, Hobart, J. Walch & Sons, 1880, pp. 63–5.

4 The island's Indigenous people were subject to near-genocide. Within 30 years of white settlement the nine tribes had been decimated through armed conflict, introduced diseases and dispersion. Billy was William Lanne, the last full-blood Aboriginal male, whose body was mutilated after death as part of a grisly conflict for possession between Tasmania's Royal Society and the Royal College of Surgeons in England. Truganini became celebrated

as the last full-blood Tasmanian Aboriginal person. She died in 1876 and her skeleton was displayed in the Tasmanian Museum for many years, then kept hidden there. The Museum returned it to the Aboriginal community in 1976 and she was finally laid to rest in a ceremony on the waters of the D'Entrecasteaux Channel. Enlightened though she was in her time, Mary Roberts' casual use of these names is a sure indicator that notions of romantic savages still beat strongly in the Empire's bosom.

5 Roberts, Mary G., 'The Keeping and Breeding of Tasmanian Devils (*Sarcophilus harrisii*)', *Proceedings of the Zoological Society of London*, 1915, pp. 1–7.

6 ibid.

7 ibid.

8 Flynn, T. T., 'Contributions to a Knowledge of the Anatomy and Development of the Marsupiala [:] No. I. The Genitalia of *Sarcophilus satanicus*', *Proceedings of the Linnean Society of New South Wales,* vol. xxxv, Part 4, 30 November 1910. [Issued 1 March 1911], p. 873.

9 ibid.

10 ibid., p. 874.

11 Guiler, Eric, 'The Beaumaris Zoo in Hobart', *Tasmanian Historical Research Association Papers and Proceedings*, vol. 33, no. 4, December 1986, p. 128.

12 Lord, Clive, 'Existing Tasmanian Marsupials', *Royal Society of Tasmania Papers & Proceedings*, Hobart, 1927, p. 22.

13 ibid., p. 24.

Chapter 6

1 *Ardmore Daily,* Ardmoreite (Oklahoma), 8 December 1927, page 8.

2 *Tavistock Gazette* (UK), Friday 2 July 1869, page 5.

3 *Western Daily Press* (UK), Tuesday 2 June 1868, page 3.

4 *Oxford Chronicle and Reading Gazette*, Saturday 12 September 1868, page 8.

5 *Irish Emerald*, Saturday 28 October 1899, page 13.
6 *Bath Chronicle and Weekly Gazette*, Thursday 14 March 1867, page 3.
7 *Whitby Gazette*, Saturday 10 February 1866, page 2.
8 *The Era* (London, UK), Sunday 21 June 1868, page 12.
9 *Ballymena Observer*, Saturday 7 May 1870, page 3.
10 *The Maitland Mercury and Hunter River General Advertiser* (NSW), Thursday 11 May 1871, page 4.
11 *Colonies and India* [London, UK], Saturday 27 November 1880, page 15.
12 *The Tasmanian* (Launceston), Saturday 25 February 1882, page 209.
13 *Daily Telegraph & Courier* (London), Saturday 20 October 1883, page 3.
14 *The Indianapolis Journal*, Sunday 16 August 1885, page 4.
15 *Los Angeles Herald*, 1 August 1909, page 15.
16 *Sydney Mail*, Wednesday 26 November 1913, page 29.
17 *Mount Alexander Mail* (Victoria), Tuesday 22 February 1881, page 4.
18 *Leicester Daily Post*, Tuesday 11 July 1911, page 5; *Washington Evening Star*, 10 September 1905, page 22.
19 *The Sydney Mail and New South Wales Advertiser*, Wednesday 2 December 1903, page 1444.
20 *The Scotsman*, Saturday 28 December 1867, page 2.
21 *The Mercury*, Monday 6 March 1876, page 2.
22 *Bexhill-on-Sea Observer* (UK), Saturday 3 January 1925, page 9.
23 *Bradford Daily Telegraph*, Thursday 27 July 1871, page 1.
24 *Saunders's News-Letter*, Thursday 5 May 1870, page 2.
25 *Daily Telegraph* (Launceston), Monday 15 June 1891, page 4.
26 *Gippsland Mercury* (Victoria), Thursday 26 June 1884, page 2; *Glasgow Herald*, Tuesday 5 February 1884, page 6 [A Tour Round the World] 'Australia. Some Strange things seen by Moncure D Conway.'

27 *Westminster Gazette*, Thursday 19 September 1907, page 5.
28 *Daily Post* (Hobart), Monday 31 May 1909, page 3.
29 *Telegraph* (Brisbane), Wednesday 16 May 1934, page 20.

Chapter 7

1 *Voice* (Hobart), Saturday 13 March 1943, page 1.
2 *Malus pumila* is the botanical name for the common apple tree. In Latin, *malus* can mean both 'apple' and 'evil'. The apple's easy conversion to cider may explain the latter.
3 McConnel, Anne and Servant, Nathalie, 'The History and Heritage of the Tasmanian Apple Industry—A Profile', Report of the Queen Museum and Art Gallery, Launceston, December 1999, page 49. (A National Estate Grants Program Funded Study.)
4 *Lyttleton Times* (UK), Wednesday 22 May 1867, page 2.
5 *Liverpool Daily Post*, Tuesday 10 May 1870, page 1.
6 *The Sydney Morning Herald*, Thursday 14 February 1861, page 1.
7 www.bie-paris.org/site/en/blog/entry/5-things-you-might-not-know-about-expo-1880-melbourne, accessed 24 June 2023. (Website of Bureau International des Expositions.)
8 *The Mercury*, Tuesday 23 November 1880, page 3; *The Mercury*, Wednesday 1 December, page 2.
9 *Airdrie & Coatbridge Advertiser*, Saturday 30 June 1906, page 3.
10 *Gloucestershire Echo*, Friday 22 February 1907, page 3.
11 *Cornwall Advertiser* (Launceston), Friday 7 June 1872, page 5.
12 *The Mercury*, Thursday 30 July 1936, page 2.
13 *Coventry Evening Telegraph* UK), Monday 4 April 1938, page 12.
14 *Belfast Telegraph*, Friday 22 October 1937, page 11.
15 *The Mercury*, Saturday 13 November 1937, page 11.
16 *Gravesend & Northfleet Standard*, Saturday 1 September 1906, page 7.
17 *Colonial Times* (Hobart), Tuesday 18 June 1833, page 2.
18 *Adelaide Observer*, Saturday 4 May 1867, page 1.
19 *The Argus* (Melbourne), Saturday 18 July 1874, page 4.

20 *Cornubian and Redruth Times* (UK), Friday 5 June 1868, page 2.
21 *Glasgow Evening Citizen* (UK), Tuesday 18 August 1868, page 2.
22 *Leicester Journal* (UK), Friday 11 September 1868, page 3.
23 *Mercury*, Saturday 17 August 1872, page 2.
24 *The Argus* (Melbourne), Friday 25 July 1873, page 1.
25 *Australian Town and Country Journal* (Sydney), Saturday 7 August 1875, page 15.
26 *Launceston Examiner*, Tuesday 27 December 1881, page 2.
27 *Launceston Examiner*, Tuesday 10 January 1882, page 3.
28 *The Sydney Mail and New South Wales Advertiser*, Saturday 17 June 1882, page 978.
29 *Strathearn Herald* (UK), Saturday 7 June 1884, page 3; *Bristol Times and Mirror*, Tuesday 24 June 1884, page 5.
30 *Launceston Examiner*, Saturday 21 February 1885, page 3.
31 *The Age*, Saturday 19 March 1892, page 3; *Traralgon Record* (Victoria), Tuesday 14 June 1892, page 2.
32 *Daily Telegraph* (Launceston), Saturday 28 July 1906, page 5.
33 *Mercury* (Hobart), Monday 19 February 1923, page 6.
34 *Glen Innes Examiner* (NSW), Saturday 25 May 1946, page 2.
35 *Advocate* (Burnie), Friday 14 June 1946, page 5.
36 *Mercury*, Thursday 13 May 1954, page 10.
37 *Port Lincoln Times* (South Australia), Thursday 6 August 1970, page 6.

Chapter 8

1 *Voice* (Hobart), Saturday 30 September 1939, page 8, quoting the *Los Angeles Times.*
2 *Werribee Shire Banner* (Victoria), Thursday 1 February 1940, page 2.
3 *Mercury*, Monday 24 October 1949, page 4; *Belfast Telegraph*, Monday 15 April 1940, page 5.
4 *Daily Telegraph* (Sydney), Sunday 8 June 1941, page 6.
5 *The Era* (UK), Sunday 8 August 1875, page 3.
6 *Evening Journal* (Adelaide), Saturday 26 May 1883, page 1.

7 *The Mercury*, Wednesday 13 June 1883, page 3.
8 *Nottinghamshire Guardian* (UK), Saturday 10 October 1896, page 7.
9 *Smethwick Telephone* (UK), Saturday 21 September 1907, page 4.
10 *Telegraph* (Brisbane), Friday 23 June 1922, page 3.
11 *Weekly Dispatch*, Sunday 18 December 1927, page 5.
12 *Hartlepool Northern Daily Mail* (UK), Thursday 7 March 1935, page 6.
13 *Mercury*, Thursday 23 August 1945, page 7.
14 *News* (Adelaide), Monday 23 April 1951, page 4.
15 Advertisement in *San Rafael Daily Independent Journal* (California), 10 November 1958, page 22.
16 *Brisbane Telegraph*, Thursday 7 August 1952, page 34.
17 *Bendigo Advertiser*, Wednesday 27 September 1899, page 6.

Chapter 9

1 Guiler, Eric, *The Enthusiastic Amateurs: The Animals and Birds Protection Board 1929–1971*, Sandy Bay, E. R. Guiler, 1999, p. 73.
2 The published results are in Guiler, E. R., 'Observations on the Tasmanian Devil, *Sarcophilus harrisii* (Dasyuridae: Marsupiala) at Granville Harbour, 1966–75', *Papers and Proceedings of the Royal Society of Tasmania*, vol. 112, 1978, Hobart, The Society, 1978, pp. 161–88. See also Guiler, E. R. and Heddle, R. W. L., 'Observations on the Tasmanian Devil, *Sarcophilus harrisii* (Dasyuridae: Marsupiala). 1. Numbers, home range, movements and food in two populations', *Australian Journal of Zoology*, 18(1), 1970, pp. 49–62.
3 *Australian Wild Life: Journal of the Wild Life Preservation Society*, vol. 3, no. 3, March 1958, Sydney, The Society, 1958, p. 14.
4 ibid.
5 *Australian Wild Life*, op. cit., vol. 4, no. 2, 1962, pp. 30–2.
6 *Australian Outdoors*, November 1961, Sydney, The Society, p. 36.
7 ibid., p. 37.
8 Known as 'Damper Inn' because it was damper in than out.

9 Bauer, Jack, 'Protection That Doesn't Protect', *Australian Outdoors*, November 1961, Sydney, The Society, pp. 36–41.

Chapter 10

1 Guiler, E. R., 'Observations on the Tasmanian Devil', p. 169.
2 ibid., p. 177.
3 ibid., p. 183.
4 *The Mercury*, 9 August 1966, p. 6. The area covered a 'fifty-mile radius' from Tooms Lake in the east to Interlaken across the Western Tiers, and south to Swansea.
5 *The Mercury*, 15 January 1972, p. 4.
6 *The Mercury*, 1 July 1972, p. 3.
7 Guiler, Eric, 'Tasmanian Devils and Agriculture', *Tasmanian Journal of Agriculture*, May 1970, p. 137.
8 *Launceston Examiner*, 28 January 1987, p. 3.
9 *Tasmanian Country*, 26 June 1987, p. 2.
10 *The Mercury*, 6 August 1975, p. 14.
11 'Tasmania. Ministerial News Release No. 1521, October 27, 1984.'
12 *The Mercury*, 2 February 1988, p. 1.
13 *The Mercury*, 16 October 1985, p. 1. Pam Clarke went on to become a leading world campaigner against the practice of battery hen egg production, for which she has an impressively long record of arrests and court appearances. In the leadup to the Sydney 2000 Olympic Games she gained considerable publicity for her campaign by saying that its official logo looked like 'a sad chook'.
14 *The Mercury*, 17 October 1985, pp. 1–2. The B.Sc. (Hons) thesis in question: 'The Cranial Anatomy and Thermoregulatory Physiology of the Tasmanian Devil, *Sarcophilus harrisii* (Marsupiala: Dasyuridae)', 1984, by Syed K. H. Shah, University of Tasmania, Hobart.
15 *The Mercury*, 7 July 1988, p. 1.
16 *The Sunday Tasmanian*, 23 July 1988, p. 5.

17 Mooney, Nick, 'The Devil you know', *Leatherwood: Tasmania's Journal of Discovery*, vol. 1, no. 3, Winter 1992, Hobart, Allan Moult, 1992, pp. 54–61.

Chapter 11

1 Anderson, Angela, interview with David Owen, 24 January 2004.
2 www.kidszoo.com, accessed 10 April 2004.
3 The interview was conducted between 7 and 9 April in 2004.
4 Email dated 19 May 2004.

Chapter 12

1 Flynn, Errol, *My Wicked, Wicked Ways*, Cutchogue, New York, Buccaneer Books, 1976. Typical of the larrikin style of the book, Errol also refers to his father as 'just a tall hunk of scholarship' (p. 19).
2 Flynn, T. T., 'Contributions to a Knowledge of the Anatomy and Development of the Marsupiala [:] No. I. The Genitalia of *Sarcophilus Satanicus*', *Proceedings of the Linnean Society of New South Wales,* vol. xxxv, Part 4, 30 November 1910. {Issued 1 March 1911}, p. 873.
3 Norman, Don, *Errol Flynn: The Tasmanian Story*, Hobart, W. N. Hurst & E. L. Metcalf, 1981, p. 4.
4 Flynn, Errol, op. cit., p. 24.
5 ibid., p. 104.
6 Jack Warner, quoted in *Hollywood Be Thy Name: The Warner Brothers Story*, by Cass Warner Sperling, Rocklin, CA, Prima, 1994, p. 195.
7 Flynn, Errol, op. cit., p. 168.
8 Warner, op. cit., p. xi.
9 ibid., p. 7 and p. 343.
10 Jones' inspiration for the coyote—a scavenging carnivore—came from an earlier creative interpretation: 'I first became interested in the Coyote while devouring Mark Twain's *Roughing It* at the age

of seven. I had heard of the coyote only in passing references from passing adults and thought of it—if I thought of it at all—as a sort of dissolute collie. As it turned out, that's just about what a coyote is, and no one saw it more clearly than Mark Twain[:] "The coyote is a long, slim, sick and sorry-looking skeleton, with a gray wolf-skin stretched over it, a tolerably bushy tail that forever sags down with a despairing expression of forsakenness and misery, a furtive and evil eye, and a long, sharp face, with slightly lifted lip and exposed teeth. He has a general slinking expression all over. The coyote is a living, breathing allegory of Want. He is *always* hungry. He is always poor, out of luck and friendless . . . He does not mind going a hundred miles to breakfast, and a hundred and fifty to dinner, because he is sure to have three or four days between meals . . .'" Jones, Chuck, *Chuck Amuck: the Life and Times of an Animated Cartoonist*, New York, Farrar, Straus & Giroux, 1989, pp. 34–5. (Twain visited the Tasmanian Museum and Art Gallery in 1897. He seemed to have difficulty identifying a Tasmanian devil and oddly referred to a highly predatory Tasmanian sheep-killing parrot that feasted only on its victims' kidney fat. He presumably meant the Kea, a scavenging carnivorous parrot found only in New Zealand.)

11 Sandler, Kevin S. (ed.), *Reading the Rabbit: Explorations in Warner Bros. Animation*, New Brunswick, NJ, Rutgers University Press, 1998, p. 7.

12 Jones, op. cit., p. 109.

13 Beck, Jerry and Friedwald, Will, *Warner Bros. Animation Art: the Characters, the Creators, the Limited Editions*, Westport, CT, Hugh Lauter Levin Associates Inc/WB Worldwide Publishing, 1997, pp. 74–5.

14 ibid., pp. 129–30.

15 www.errolflynn.net/Filmography, accessed 30 December 2003.

16 Bevilacqua, Simon, *Sunday Tasmanian*, 10 May 1998, p. 7.

17 Taz looks not unlike a very young devil, which has a disproportionately big head and tucked-in, obscure limbs.

18 Lenburg, Jeff, *The Encyclopedia of Animated Cartoons*, 2nd edn, New York, Facts on File, 1999, p. 142.
19 Grant, John, *Masters of Animation*, London, BT Batsford, p. 154.
20 Jones, op. cit., pp. 92, 93.

Chapter 13

1 McCorry, Kevin, http://looney.toonzone.net/articles/tazarticle.html, accessed 14 June 2004.
2 Sandler, Kevin S. (ed.), *Reading the Rabbit: Explorations in Warner Bros. Animation*, New Brunswick NJ, Rutgers University Press, 1998, p. 177.
3 Bevilacqua, Simon, *Sunday Tasmanian*, 28 September 1997.
4 ibid., p. 6.
5 ibid., p. 7.
6 ibid., pp. 1, 6.
7 ibid., p. 6.
8 ibid., p. 6.
9 *Sunday Tasmanian*, 5 October 1997, p. 3.
10 ibid., pp. 14, 15, 45.
11 *Hansard*, 15 October 1997. Hobart, Parliament of Tasmania, October 1997.
12 *Sunday Tasmanian*, 10 May 1998, p. 3.
13 *Hansard*, 15 October 1997.

Chapter 14

1 Quammen, David, *Spillover: Animal Infections and the Next Human Pandemic*, London, Vintage Books, 2012, back cover.
2 Patchett, Amanda, et al., 'Two of a Kind: Transmissible Schwann Cell Cancers in the Endangered Tasmanian Devil (*Sarcophilus harrisii*), *Cellular and Molecular Life Sciences*, 2020, vol. 77, no. 9, pp. 1847–58.
3 Fox, Samantha, et al., 'The Road to Recovery: A Recipe for Success?' in *Saving the Tasmanian Devil: Recovery through Science-based Management*, 2019, CSIRO Publishing, p. 273.

4 'Tasmanian Devils Set Free in Military Training Area', *Royal Australian Navy News*, Thursday 22 September 2016, p. 2; *Army*, Thursday 22 September 2016, p. 8.
5 ibid.
6 Lyons, Bruce A., and Woods, Gregory M., 'Immune Strategies to Combat DFTD', in *Saving the Tasmanian Devil: Recovery through Science-based Management*, pp. 77–82.
7 Biggs, James R., et al., 'Advocates and Ambassadors: The Devil Is Real', in *Saving the Tasmanian Devil: Recovery through Science-based Management*, p. 201.
8 'Tasmanian Devil Ambassador Program in Europe', www.zoochat.com/community/threads/tasmanian-devil-ambassador-program-in-europe.475 605/, accessed 11 July 2023; 'Tasmanian Devil Ambassador Program in Europe', www.zoochat.com/community/threads/tasmanian-devil-ambassador-program-in-europe.475 605/page-3, accessed 11 July 2023.
9 ibid.
10 Irons, Greg, email to David Owen, 13 July 2023.
11 ibid.
12 'What Is Rewilding?' https://rewildingeurope.com/what-is-rewilding/, accessed 2 July 2023.
13 Fox, Samantha, and Seddon, Philip J., 'Wild Devil Recovery: Managing Devils in the Presence Of Disease', in *Saving the Tasmanian Devil: Recovery through Science-based Management*, p. 158.
14 Dahlstrom, Michael, Environment Editor, *Yahoo News Australia*, 1 February 2023, https://au.news.yahoo.com/200-deaths-in-just-two-years-devastating-toll-of-deadly-aussie-road-tasmanian-devil-062439588.html, accessed 1 June 2023.
15 'John F. Eisenberg, 68, Dies; Leading Expert on Mammals', www.nytimes.com/2003/07/20/us/john-f-eisenberg-68-dies-leading-expert-on-mammals.html, 20 July 2003.

SELECT BIBLIOGRAPHY

Beck, Jerry, and Friedwald, Will, *Warner Bros. Animation Art: the Characters, the Creators, the Limited Editions*, Westport, CT, Hugh Lauter Levin Associates Inc/WB Worldwide Publishing, 1997.

Brogden, Stanley, *Tasmanian Journey*, Melbourne, Morris & Walker for Pioneer Tours, 1948.

Eisenberg, John, *The Mammalian Radiations*, Chicago, University of Chicago Press, 1986.

Ewer, R. F., *The Carnivores*, London, Weidenfeld and Nicolson, 1973.

Farrand, John Jr., ed., *The Audubon Society Encyclopedia of Animal Life*, New York, Chanticleer Press, 1987 [Sixth Printing, 1988].

Flood, Josephine, *Archaeology of the Dreamtime*, Sydney, Collins, 1983.

Flynn, Errol, *My Wicked, Wicked Ways,* Cutchogue, New York, Buccaneer Books, 1976.

Gilbert, Bill, *In God's Countries*, Omaha, University of Nebraska Press, 1984.

Gould, John, *The Best of Gould's Mammals: Selections from* Mammals of Australia Volumes l, ll and lll, selected and introduced with modern commentaries by Joan Dixon, South Melbourne, Macmillan, 1977 (rev. edn 1984).

Grant, John, *Masters of Animation*, London, B. T. Batsford, 2001.

Green, R. H., *The Mammals of Tasmania*, Launceston, Foot & Playsted, 1973.

Grzimek, Bernhard, *Australians{:} Adventures with Animals and Men in Australia*, Translated by J. Maxwell Brownjohn, London-Sydney, Collins, 1967.

Guiler, Eric, *The Enthusiastic Amateurs: The Animals and Birds Protection Board 1929-1971*, Sandy Bay, E. R. Guiler, 1999.

Guiler, Eric, *Marsupials of Tasmania*, Hobart, Tasmanian Museum and Art Gallery, 1960.

Guiler, Eric, *The Tasmanian Devil*, Hobart, St David's Park Publishing, 1992.

Jones, Chuck, *Chuck Amuck: the Life and Times of an Animated Cartoonist*, New York, Farrar Straus Giroux, 1989.

Jones, Menna, Dickman, Chris, Archer, Mike, eds, *Predators with Pouches: the Biology of Carnivorous Marsupials*, Collingwood, Vic., CSIRO, 2003.

Le Soeuf, W. H. Dudley, *Wildlife in Australia*, Christchurch [NZ], Melbourne, Whitcombe and Tombs, 1907.

Lenburg, Jeff, *The Encyclopedia of Animated Cartoons*, 2nd ed., New York, Facts on File, 1999.

Long, John, Archer, Michael, Flannery, Timothy, Hand, Suzanne, *Prehistoric Mammals of Australia and New Guinea: One Hundred Million Years of Evolution*, Baltimore, Johns Hopkins University Press, 2002.

Lord, Clive E., and Scott, Herbert Hedley, *A Synopsis of the Vertebrate Animals of Tasmania,* London, Oldham, Hobart, Beddome and Meredith, 1924.

Meredith, Louisa Anne, *My Home in Tasmania, During a Residency of Nine Years*, London, John Murray, 1852.

Meredith, Louisa Anne, *Tasmanian Friends and Foes: Feathered, Furred and Finned; A Family Chronicle of Country Life, Natural History, and Veritable Adventure*, Hobart, J Walch & Sons, 1880.

Mooney, Nick, 'The Devil you know', in *Leatherwood: Tasmania's Journal of Discovery*, Volume 1 Number 3, Winter 1992, Hobart, Allan Moult, 1992.

Morrison, Reg, and Morrison, Maggie, *The Voyage of the Great Southern Ark*, Sydney, Lansdowne Press, 1988.

Norman, Don, *Errol Flynn: The Tasmanian Story*, Hobart, W. N. Hurst and E. L. Metcalf, 1981.

Nowak, Ronald M., *Walker's Mammals of the World*, Baltimore, Johns Hopkins University Press, 1999.

Sandler, Kevin S., ed., *Reading the Rabbit: Explorations in Warner Bros. Animation,* New Brunswick, NJ, Rutgers University Press, 1998.

Sperling, Cass Warner, *Hollywood Be Thy Name: the Warner Brothers Story*, Rocklin, CA, Prima, 1994.

Strahan, Ronald, ed., *The Mammals of Australia*, rev. edn, Chatswood, Reed Books, 1995.

Taylor, James, comp., *Zoo: Studies From Nature*, Sydney, James Taylor, 1920.

Watts, Dave, *Tasmanian Mammals: A Field Guide*, Hobart, Tasmanian Conservation Trust, 1987.

Willoughby, Howard, *Australian Pictures Drawn With Pen and Pencil*, London, The Religious Tract Society, 1886.

Wroe, Stephen, 'The Myth of Reptilian Domination', in *Nature Australia*, Summer 2003-2004, p. 59.

www.wolverinefoundation.org

www.rokebyprimary.tased.edu.au/NAIDOC

www.napak.com/honey_badger

www.awf.org/wildlives

www.environment.sa.gov.au/parks/naracoorte

www.jonacohen.com/jersey_devil

www.errolflynn.net/Filmography

www.kidszoo.com

INDEX

ABC Radio *PM* 23
Aboriginal people, Tasmanian
  diet, speculation as to 38–9
  hunters, as 38
  Marrawah, significance to 1
Age of Mammals 34
AIF (Australian Imperial Forces)
  12th Battalion, devil mascot 119
AIF Originals 121
Albion, Ingrid 1
Allport, Curzon 5
Allport, Morton 5–6
Anderson, Angela 159–60, 161, 162, 163
Animal Liberationists 152, 153
Animals and Birds Protection Board 110, 126
  export of devils 128–30
antechinus 56–7, *56*
Anti-Plumage League 80
apple industry 107–8
Archer, Michael 31
Arthur Lakes 139
Arthur River 19
Attenborough, Sir David 32
Australia, mainland 6, 30
  extinction of devil on 37, 209
Australian Army
  12th/40th Royal Tasmanian Regiment mascot 201–2
Australian Journal of Zoology 154
Australian Museum 32
Australian Outdoors 134
Australian Wild Life 129, *129*, 130–4
Avery, Tex 177

Baars, Christo 3
Baker, H. D. 102
Basel Zoological Gardens 129
Bauer, Jack 134–5, 145, 158, 207
  'Protection That Doesn't Protect' 135–44
bears 10
Beaumaris Zoo 19, 65, 80, 82, *84*, 89
Beelzebub 6

Beelzebub's pup 6
Ben Lomond (mountain) 9
Bigg, Keith and Colleen 50
Birds of Australia 77
blitzkrieg hypothesis 36
'Bluey' mascot 201–2, *202*
Bonaparte, Joseph 16
Bonaparte, Josephine 16
Bonorong Wildlife Sanctuary 25, 160, 206
Bostock and Wombwell travelling menagerie 109–10
Bothwell 81
Boviak Beach 6, 8
Buckland 20
Buckland and Spring Bay Tiger and Eagle Extermination Society 58
Bugs Bunny 176, 177, 178
Burnie 110
Butchart, Robert 200

Cairo Zoo 122, 201
Cameron, Aunty Patsy XII
Cameron, Major R. 9
Cameron, Max 17
cancer and 2
  transmissible forms 195
Cape Barren geese 203
Cape Portland 128
Captain Blood 174
captivity 82–9, 100–3, 105, 110–11, 159
  breeding, captive 63–5, 84–6, 87, 161
  breeding in overseas zoos, prohibition on 205
  escape stories 112–18
  Fort Wayne Children's Zoo 164–8
  Tasmanian Devil Park (Unzoo) 159–63
  zoos, display in 99–103, 105, 110–11, 159
Carlton and United Breweries 'Save Our Native Fauna' series *210*
carnivore guild 53, 211–12
Carson, Alice 211
cats 10
cattle and 3, 7, 11, 12
Clampett, Bob 177
Clarke, Pam 152–3
Clifford, Bob 187
climate change 35
Coleman, Donna 59
Coles Bay 12
Colvin, Mark 23
commercial exploitation of devils 100–1, 105–11
convergent evolution 44–50
convicts and 96
Conway, Moncure D. 101
Copenhagen Zoo 205
Cradle Mountain 20, 139, 155
Cronin, Steve 61
Croome, Rodney 74
Crotty Track 139

CSIRO (Commonwealth Scientific and Industrial Research Organisation) 94
Cuvier, Georges 7

Daintree rainforest 36
Damper Inn 140
Darling Downs 33
Dasyuridae 52, 67
Davis, George 20, 24–5, 26
Department of Defence 200
Derwent River 5
devil facial tumour disease (DFTD) 1, 2, 9, 37, 43, 94, 161
  arrival of 194
  avoidance of extinction caused by 95, 196
  birth time, changes to 62
  clinical definition 3
  death after mating 56
  DFT1 195
  DFT2 195
  first official case, description 3–4
  Forestier Peninsula project 199, 203
  incurable 194
  insurance population 197, 198, 199, 204
  Maria Island program 199–200
  no historical account of 26
  population, impact on 196
  Schwann cell 196
  spread of, speed 197
  transmission, method of 195
  treatments 203–4
Devil May Hare 180–1
Devil Tracker Adventure project 163
Devil's Camp 25
Devil's Lair Cave, WA 33, 38
Diabolus ursinus 7
Didelphis ursina 75
dingo 37, 38, *42*
dogs, feral 7
Downie, R. J 147
Dunalley 163, 203
dwarfism 36, 59

eagle, wedge-tailed 57–8, *58*, 73
eastern Darling Downs fossil site 33
Eisenberg, John 53, 211
  Carnivore Park, call for establishment of 211–12
ENSO effect 35–6
equal spacing 54–5
Ewer, R. F. 43
Examiner (Launceston) 105–6, 148
export of devils 81, 82, 105, 108–11
  Animals and Birds Protection Board involvement 128–30
extinctions 31
  antipodean mass 35
  thylacine 40

farmers 24–5, 26, 134–5, 136, 147–50
fire management 36
Fishermans Cliff, NSW 33
Fleay, David 17, 18, 19, 25
  captive breeding 63–5
Flinders Island 12
Flood, Josephine 39
Flynn, Errol 90, 92, 169, 170–2, 173, 174–6, *175*, 178, 182
Flynn, Lily (Marelle) 170–3
Flynn, Theodore Thomson 86, 89–92, 152, 169, 170–2, 173, 182
Forestier Peninsula project 199, 203
Fort Wayne Children's Zoo, US 164
fossils 7, 29, 31–2, 59
fox 10, 50, 157
Fox, Samantha 200
Frankfurt Zoo 18
Freling, Friz 177
Freycinet Peninsula 10, 12
Furred Animals of Australia 134, 137

Gales, Rosemary 22
Gee, Helen 20
George, Brian 28
Giles, Barry 189
Glaucodon ballaratensis 31
goanna 34
Godthelp, Henk 31
Gondwana 29
Gould, John 77–8, 80
Granville Harbour 128, 145
Grey, Lionel 20
Groom, Ray 152, 187–8, 190–2
Grzimek, Bernhard 18–19
Guiler Eric 126–8, *128*, 145–6, 169
  Animals and Birds Protection Board 126
  bounty records 7
  devil feeding 11, 135, 136
  devil handling 13
  devil hermaphroditism 51
  devil mortality 65
  devil population fluctuations 146–8
  devil sheep predation 127
  devil singing 19
  devil speed 19
  devil surveys 128, 146
  devil traps 17
  Mary Roberts and TT Flynn, comments on 92
  The Tasmanian Devil 156

habitat fragmentation 209
Hall, Lois 149–50
Hand, Suzanne 31
Harris, General George Prideaux 7, 18, 74–5, 80, 93
  devil sketch 76
  thylacine sketch 76
Hendra virus 209

Hibbert, Kate 200–1
Hobart Walking Club 140
Holdsworth, Mark 8
Holocene Epoch 33, 36
honey badger *see* ratel
Hood, Geoff *140*
Hooper, Reuben 19
human interference theory 36
hunting, uncoordinated social 9
hyaena 10, 22, *42*, 45, 50–2, *51*

Ice Age 32, 34, 35
Incat Tasmania 187
Indigenous names for the Tasmanian devil 2
International Union for the Conservation of Nature (IUCN) Red List 206
  Endangered listing 197
Irons, Greg 160, 206–8

Jersey Devil, legend of 16, 17
Johnstone, Maureen 126
Jones, Chuck 177, 181
Jones, Menna 3, 12
  devil breeding 61
  devil competition with quolls 53–4
  devil dentition 41, 42, 53
  devil DFTD 3
  devil speed 20

kangaroo 38
Kelly, Androo 167
King, Geoff 1–2, 10, 212
King Island 12
King's Run 1
Kirchner, Elaine 164–8
koala 152
kowari 57

Lake Nitchie, NSW 39
Lake St Clair 127, 139
Lang, Dr E. M. 129, 130
Leatherwood: Tasmania's Journal of Discovery 156
Linnean Society of London 76
lion 28
lion hound 175
Little Swanport 3
London Zoological Society 7, 77, 81, 100–1
Looney Tunes 92, 176, 177, 178
Lord, Clive 15, 44, 89, 91, 93
Los Angeles Zoo 121

McCrossen, Brendan 40
McKimson, Chuck 178–80
McKimson, Robert (Bob) 169, 177, 178–80
Macquarie University 198
Mammalian Radiations, The 53
Mammals of Australia 77
Mammoth Cave, WA 33
Marcus, Sid 178
Maria Island program 199–200, 204
  prey on other animals of 202–3
Marrawah 1, 212

marsupial carnivore xi, 4, 16, 28
media reports about 122–5
megafauna 32, 33–5, 36, 55
Megalania prisca 33–4
Melbourne 74
Melbourne International
    Exhibition, 1880 108
Melbourne Zoo 18, 102–3
Menzies, Robert 183
Menzies Institute for Medical
    Research 197–8
The Mercury 108–9, 130, 152
Meredith, Louisa Anne 78–80
Merrie Melodies 177
Miocene Epoch 31
Monarto Zoo (SA) 200
Mooney, Kate 13, 14, 22
Mooney, Nick 3, 9, 25, 51,155,
    155
  devil behaviour 14
  devil dens 21–2
  DFTD 2
  devil feeding 8, 10, 157
  devil latrines 22
  devil scats 157
  devil swimming ability 19
  hand-rearing of orphans 13–14,
    22, 23
  'The Devil You Know' 156–8
Morita, Annie 185
moths 9
Mount Wellington 186
Mount William (wukalina) 3, 61,
    200
Muller, Prof Konrad 153
Musselroe Bay 73
My Wicked, Wicked Ways 171

Naracoorte Caves, SA 32
Narawntapu National Park 10
Nature Conservation Branch 8
New Holland 6
Nicol, S. C. 153, 154
Noetling, Fritz 38–9
Nurse, Jenny 145
Nutella (devil) 200–1, *201*

Oatlands 147
Obendorf, Dr David 156
Olympic Games, Sydney 2000
    185–6, 191
opossum, American 75
Osaka Zoo 151
Owen, David 164–8
Owen, Richard 7, 33

Pacific Ocean 35
Parer, David 12
Parks and Wildlife 25
Pearse, D. Colbron 15, *15*
Pelham 26
Pemberton, David 14, 20, 23, 46,
    51, 61, 62, 211
  DFTD 3
  devil feeding 10, 68
  devil feeding postures 69–71
  devil home range 73
  devil vocalisations 69

hand-rearing orphans 13, 22, 23
'Social Organisation and Behaviour of the Tasmanian Devil, *Sarcophilus harrisii*' 156
penguins 202–3
Perry, Richard 159
phascogale 57
planigale 57
Pleistocene Epoch 32
Pliocene Epoch 31
polyphagous carnivores 20
Port Arthur Historic Site 160
Port Phillip Bay 74
possum 10, 16, 38, 48
poultry 17
Prague Zoo 204, 206
Putt, Peg 192

Quammen, David 195, 209
Queen Victoria Museum and Art Gallery 33
Quinkana fortirostrum 34
quoll, eastern 23, 52, 54, 55, 211
quoll, spotted-tailed 10, 15, 16–17, 52–4, *54*, *55*, 57, 66, 73, 211
quoll genus 31, 35, 36, 52–3

rabbit 9, 16
Randall, David 10, 19, 26
ratel 22, *42*, 45, 48–50, *49*
Mellivora capensis 45
Reece, Eric *128*
rewilding 208–9
Richter, Henry 77
Riversleigh fossil deposits 31–2
roadkill *11*, 12, *26*, 161, 162, 203, 211
Roberts, Mary 19, 66, 80–9, *82*, 92, 145, 158
Rowell, Marcus 188
Royal Society of Tasmania 38

Sadler, Debbie 5
Saint-Hilaire, Geoffroy 7
San Diego Zoo 22, 204
Sapporo Maruyama Zoo 151–2
Sarcophilus harrisii 7, 33, *37*
Sarcophilus laniarius 7, 32, 33
Sarcophilus moornaensis 7
Sarcophilus satanicus 7, 91
Sarcophilus ursinus 112, 172
satanic tag 98
Save the Tasmanian Devil Program 13, 196–7, 198, 200
Saving the Tasmanian Devil: Recovery Through Science-Based Management 94
Scott, Alan 9, 11
Selzer, Eddie 181
semelparousness 56
sheep and 3, 7, 9, 12, 20, 26, 48, 107, 136, 157
Smithsonian Institution 16
snaring 16, 25, 89
snow leopard *42*
'Social Organisation and Behaviour of the Tasmanian Devil, Sarcophilus harrisii' 156

Spielberg, Steven 184
Spillover: Animal Infections and the Next Human Pandemic 195
Stead, Thistle Y. 130–4
Sunday Tasmanian 178, 187
Sutton, Garry 10

Table Mountain 139
tadpoles 9
tapeworms 4, 12
Tarkine/takayna 212
Tasman Peninsula 163
  DFTD-free, efforts to keep 163
Tasmania
  Gondwanan remnant forests 36
  human population 104
  northeast, decline in devil numbers 3
  old growth forests 192
Tasmania Police 8
Tasmanian Country 149–50
Tasmanian devil 8, *15*, 36, *42*, *71*, *82*, *160*, *179*
  Aboriginal people and 38–9
  aggression 13, 15, 68–9
  birth size *61*, 61–2
  bone consumption 41, 43
  bounty 16, 48
  breeding 60–1, 62
  breeding, captive 63–5, 84–6, 87, 161, 198
  captivity *see* captivity
  claws 60
  climbing ability 47, 65, *66*, 87, 88
  commercial exploitation of 100–1, 105–11
  communal feeding 11, 12, 65–6
  cruelty towards 98–9
  cultural influence 39
  'demonising' of 96–9
  dens 21–2, 41, 71–2
  dispersal 65
  distribution 49
  Endangered listing on IUCN Red List 197
  evolution 32–5
  export of 81, 82, 105, 108–11
  extinction 35, 37
  eyesight 46, 67
  facial tumours pre-DFTD emergence 3
  fear of 8–9, 74
  feeding 1, 9, 10, *11*, 50, 60, 67–8
  feeding postures 69–71
  food source, as 38–9, 41
  food supply 27, 37, 72, 147–8
  fossils 32–3, 59
  gait *29*, 30, 49, 60
  genetic dispersal 27
  genetic diversity, low 37, 43, 194
  habitat 72
  hermaphrodites 51
  historical descriptions of 97
  home range 27, 71, 72–3, 209
  hunting 9
  hyaena, comparison with 50–2

hygienists, role as 12
illegal trade in 23–4
interaction postures 69–71
jaw gape 43
jaw strength 9, 17, 42
jewellery, teeth used in 39
juveniles *14*, 21, 23, 62, 65
killing by humans 24–5
latrines, communal 22, 70, *70*
lifespan 62
literature 71
markings 43
mating 60–1
misconceptions about 7
myths 8, 9
name 6, 7, 95–6
nocturnal 6, 47
orphaned 22, 23, 161, 162, *162*
persecution 7, 58, 94
pets 23, 141
physical description 46, 60, 75–7
play 12, 14, 23, 65, 87
poisoning 16, 24, 25, 158
population, 1990s 3
population fluctuations 25–7, 109, 146–50
pouch 51, 62, 63–4, 141
prey 9, 10, 41
protected species 23–4, 134
quoll, spotted-tailed 10, 15, 16–17
ratel, comparison with 48–50
roadkilled *11*, 12, *26*, 161, 162, 203, 211
scats 23, 157, *158*
scavenger 6, 10, 31, 34, 36–7, 40–1, 44, 50, 67
scent 17
skull 46–7, *46*
smell, sense of 46, 60
social nature 12
solitary 47, 65, 73
speed 19–20
stamina 20
sunbaking 88
tail 23, 43, 63
taxonomy 7, 52
teeth 9, 23, 39, 41–2, 60
territory, marking 47
Threatened species listing 197
thylacine and *see* thylacine
timidity 12, 13, 14, 47
trapping 25
vocalisations 6, 11–12, 18–19, 66, 69
vulnerability 2, 37
water, enjoyment of 19, 88
weight 28, 41, 55
whiskers 60
wolverine, comparison with 45–8
zoos, display in 99–103, 105, 110–11, 159
Tasmanian Devil Ambassador Program 198–9, 204

Tasmanian Devil Conservation Project 163
Tasmanian Devil Park (Unzoo) 159–63
Tasmanian emu 16
Tasmanian Greens 192
Tasmanian Museum and Art Gallery 15, 46–7, 89, 93, 172
Tasmanian Parks & Wildlife Service
  official logo *150*
Tasmanian Stockowners and Orchardists Association 147
Tasmanian Tourist Authority 150–1
Taz 13, 24, 92, 161, 176, 177, 178–80, *179*, 183–4, *191*
  cartoons made 1957–1964 181–2
  Devil May Hare 180–1
  international popularity 185
  Taz of Taz-Mania 184
Teasdale, John 183
Telegraph (Brisbane) 10
Termite Terrace 177, 181
Thompson, Bluey 119, 121–2, 201
Thomson, Donald 183
Threatened species listing 197
thylacine 7, 9, 18, 23, 31, 34–5, 36, 40, *42*, 73
  anatomic comparison with devils 41, 43, 44
  devils, relationship with 41–2, 44
Thylacine: the Tragedy of the Tasmanian Tiger 128
Thylacoleo carnifex 34
tiger 17, 28
Toledo Zoo 204
Tourism Tasmania 186
Tower Hill Beach kitchen middens, Vic 38
Trichinella pseudospiralis 156
Trichinella spiralis 155–6
Troughton, Ellis 134, 137
Trowunna 17
Trowunna Wildlife Park 167
Truganini 84
Truganini (devil) 84–7
Twain, Mark 177

University of Cambridge (UK) 198
University of Sydney 198
University of Tasmania 19, 86, 90, 152–4, 198
Unzoo (formerly Tasmanian Devil Park) 159–63

Van Diemen's Land 74, 99
Victoria Cave, SA 33

wallaby 9, 16, 27, 38, *42*, 48
Warner Bros. 10, 13, 24, 92, 119, 120–1, 170, 172, 174

animated animal characters 176–7
Taz of Taz-Mania 184
trade mark of 'Tasmanian Devil' controversy 186–92
Warner Bros. Animation Art 177–8
Warner, Jack 169, 170, 171, 174, 181, 182
Warner, Jack Jr. 175–6
water rats 10
Wayne, General Anthony 163
Weekly Times (Melbourne *210*
Wellington, Craig 188–9
Wellington Caves, NSW 33
Wheeldon, Russell 169
Wigston, Stuart 186–7
Wigston's Lures 186–7
Wild Devil Recovery Project 208–9
Wild Life Preservation Society of Australia 129–34
wildlife sanctuary 1
Williams, John 189
Wilson, John R. 104
wolf 28
wolverine *42*, *45*, 45–8, 72
Gulo gulo 45
wombat 9, 18, 25, 41, 42, 203
Wonambi naracoortensis 34
Woolnorth 211
World Heritage areas 32, 192
Wroe, Stephen 34, 36
Wynyard 173
Wynyardia bassinia 173

yabbies 9

Zeehan 146
Zoo and Aquarium Association Australasia (ZAA) 197
welfare accredited zoos *198*
Zoo Chat 205–6
zoos
breeding in overseas, prohibition on 205
display in 99–103, 105, 110–11
Tasmanian Devil Ambassador Program 198–9, 204–5
Zoos Victoria 198
Zygomatrurus 173